IRRIGATION

FOR THE

FARM, GARDEN, AND ORCHARD.

BY

HENRY STEWART,

CIVIL AND MINING ENGINEER, MEMBER OF THE CIVIL ENGINEERS' CLUB OF THE NORTH-WEST, ASSOCIATE EDITOR OF THE AMERICAN AGRICULTURIST.

WITH NUMEROUS ILLUSTRATIONS.

NEW YORK:
ORANGE JUDD COMPANY,
245 BROADWAY.

TABLE OF CONTENTS.

LIST OF WORKS AND PERIODICALS,

WHICH HAVE BEEN CONSULTED OR QUOTED IN THE PREPARATION OF THIS WORK.

Etudes sur les irrigations de Pyrenees Orientales. M. Vigan.
Economie Rurale. Boussingault.
Experiences sur l'emploi des eaux dans les irrigations. Hervé Mangon.
Italian Irrigation. M. Baird Smith.
Irrigation in Southern Europe. C. C. Scott Moncrieff.
Des Irrigations du Piemont, etc. A. Vignotti.
Manual of Hydrology. N. Beardmore.
Etude sur le service Hydraulique. De Passy.
Irrigations du midi de l'Espagne. M. Aymard.
Spanish Irrigation. C. R. Markham.
La Science des Fontaines. Dumas.
Hydrologie Agricole. De Buffon.
Traité d'hydrauliques Agricoles. Duponchel.
Outlines of Modern Farming. R. Scott Burns.
Hydraulic Engineering. C. R. Burnell.
Hydraulic Tables. John Neville.
Drainage, Irrigations, etc. M. Barrol.
Manuel de l'Irrigateur. M. Villeroy.
Irrigations et assainissement des terres. Pareto.
Theoriques et Pratiques sur les irrigations. De Cossigny.
Pratique des Irrigations en France et en Algerié. F. Vidalin.
Culture des plantes Industrielles. G. Heuzé.
Reclamation and Improvement of Agricultural Land. D. Stevenson.
Journal d'Agriculture pratique. Paris.
Reports of the Department of Agriculture. Washington.
Pacific Rural Press. San Francisco.
American Agriculturist. New York.

PREFACE.

This work is respectfully offered to those American Farmers, and other cultivators of the soil, who from painful experience can readily appreciate the losses which result from the scarcity of water at critical periods, as well as to those enterprising pioneers whose efforts are showing it to be possible to reclaim from sterility the so-called "Great American Desert." Being the first effort in American literature in this direction, it is presented with diffidence as a necessarily imperfect attempt to supply the information anxiously sought for by the classes of persons above indicated. The information here given has been acquired during some years of observation and study, and to some extent from rude, but effective practice. Hitherto there has been no work in the English language in which the practical details of irrigation were comprehensively described. The literature of the subject consists chiefly of the writings of French and Italian authors. Systems of irrigation are neither few nor modern, but all the existing methods have been the slow growth of hundreds or thousands of years. No one would be so rash as to assert that no advance upon the ancient, or even the modern methods, can be made by Americans. A nation that has done so much in originating and developing steam navigation, railroads, and the electric telegraph, can not fail to learn the best methods of utilizing streams of water, or of storing the excessive rainfall of one part of the year for use in a season of drouth. The fact that land, which without water has no salable value, immediately finds purchasers at $50 to $200 per acre, when a supply of water has been brought to it, can not fail to call attention to irrigation and quicken enterprise in its practice.

The endeavor to popularize information upon this subject and hasten the benefits that may result from its practice, is a work not without value or honor, even if it may have no other effect than to point out the way which some more competent laborer may follow with greater success.

The present is a favorable time for such an attempt. At the time these lines are written the expected seasonable rains are withheld in California, and the harvest of 1877 is endangered. Prices of wheat are rising; numerous flocks are in danger for want of water, and distress is threatened in many quarters. Farmers and gardeners in the East, who for some years past, for want of rain, have seen the profit of their labors lost, are seeking some cheap and effective methods of irrigation, by which such calamity may be avoided in future. In the great West millions of fertile acres are waiting to be reclaimed from aridity, that they may furnish homes for enterprising young men, and help to supply Europe with bread.

Irrigation is the only resource which can provide for all these contingencies. To show what has already been done; how it has been done; how our circumstances match with or differ from those in which the older systems have been applied; to offer plans and suggest cautions which occur to one who is both a farmer and an engineer, are the aims of the following pages. That the work may be at least of some help to his fellow laborers —the agriculturists of America—is the earnest wish of

THE AUTHOR.

Hackensack, N. J., January, 1877.

IRRIGATION

FOR THE FARM, GARDEN AND ORCHARD.

CHAPTER I.

THE NECESSITY FOR IRRIGATION.

The American climate is especially subject to destructive drouths, and scarcely a year passes in which the crops do not partially or wholly fail over extensive districts. That famines do not occur is not that there are no failures of crops sufficiently serious to cause them, but that our social system is so instantly helpful in case of need, that the want and misery that would otherwise certainly occur are averted by immediate and generous relief. The farmer, when rain fails, is helpless, yet there may be abundant water flowing uselessly past his suffering crops. We possess vast districts, the soil of which is of the highest fertility, but which remain barren and desert because the climate is rainless, yet large rivers flow through these arid tracts, and exhaustless subterranean streams pass through the subsoil. Water only is needed to make these tracts highly productive. The proof of this exists in the fact that already several successful efforts have been made to reclaim portions of these dry wastes by the application of a system of irrigation. But it is not only a question whether or not crops can be produced where they are now impossible, or whether or not the effects of drouths may be averted by irrigation, but whether or not the general average of the crops may be largely increased by the systematic use of partial irrigation, and the use of such supplies of water as a majority of farmers can readily avail themselves of in every part of the country.

What farmer is there who has not, in the majority of seasons, felt that some at least of his crops could have been largely benefited and increased by a copious supply of water at critical times? Market gardeners, whose crops on the average reach a value of several hundred dollars per acre, and to whom a loss of crop is partial or complete ruin, every year experience a vast amount of loss which might have been avoided were a supply of water available. A portion of this loss, in the shape of higher prices, necessarily falls upon the consumers, whose resources are insufficient to meet the increased demand; and the poorer of them are compelled in consequence to deny themselves those articles of food which are necessary to their complete health. The failure is then a public calamity. The season of 1874 was especially disastrous to strawberry growers, whose crops failed for want of rain at the season when the fruit is formed. Here were losses approaching in many cases the large sum of a thousand dollars per acre to the growers, which might have been avoided by the timely application of water. Every year there are more or less of such cases in connection with such special crops. The present year (1876) has been equally disastrous to gardeners and market farmers over a large extent in the East. The great difficulty experienced by the orange growers of Florida is precisely this want of water at critical periods. It is unnecessary to multiply instances.

No one doubts the absolute necessity of water to the growth of plants. The value of water as a nutriment or as a means of conveying nutriment to plants, however, depends upon some facts in vegetable physiology that are not generally known or considered. These may be condensed into the following statement:

Growing plants contain from 70 to 95 per cent of water. To the extent that water supplies this necessary constituent of a growing plant, it is an actual nutriment.

The solid portion of the plant consists of matters which enter into it only while in solution in water. Water is the vehicle by which the solid part of a plant is carried into its circulation for assimilation. If water is not adequately supplied, an insufficient quantity of nutriment only will be carried into the circulation of the plant, and its growth will be stunted or arrested altogether.

No water, whether it be in the state of liquid or vapor, can enter into any other part of a plant than its roots. The common idea that water or watery vapor is ever absorbed through the leaves of a plant is unfounded.

The quantity of water that must pass through the roots of a plant of our ordinary farm crops, and to be transpired through the leaves, to carry it from germination to maturity, is equal to a depth of 12 inches over the whole soil covered by the crop. This is the requirement of an average crop upon a moderately well-cultivated soil. If the crop is stimulated to extraordinary growth by large applications of manure or other fertilizers, a still greater supply of water is needed to meet the demands of the crop. Thus the yield of a crop depends in certain cases entirely upon the amount of water supplied, and to a certain extent bears an exact ratio with it.

The summer rainfall in our climate is rarely, if ever, adequate to the requirements of what would be a maximum crop, consistent with the possibilities of the soil. Our intense heats cause a large proportion of the rain-fall to be evaporated directly from the soil. Our copious summer rains are seldom wholly retained by the soil, but frequently in large part escape into streams and watercourses, and are lost to vegetation. Our fall, winter, and early spring rains come at times when the crops derive the least benefit, or none at all, from them. The amount of rain-fall that thus escapes paying tribute to our crops is by far the largest portion of it. To estimate it at three-fourths of the whole would not be unreasonable. There would then be left less than 12 inches of water to meet the necessities of the growing crops. That this sufficiently accounts for the low average of our yearly production of grass and grain is not at all improbable. The supply of water then becomes the measure of the fertility of our soil, and our climate, subject to torrid drouths in the midst of the growing season, is the obstacle to success which meets the farmer rather than the impoverished soil —a condition, indeed mainly due to a poverty of water.

To remove this obstacle to successful cultivation, it is only necessary that a system of irrigation be adopted. An adequate supply of water, ready for use in case of emergency, will render the farmer, the gardener, or the fruit grower, to a very large extent, independent of the vicissitudes of the season, and secure, beyond accident, a full reward for his labor. If with a system of irrigation a proper system of drainage be also adopted, the cultivator of the soil will have removed two adverse influences, against which he is now called upon so frequently, and so ineffectually, to strive. To irrigate economically, and successfully, however, is a business which requires a large amount of technical knowledge and skill, and the expenditure of a considerable amount of capital either in money or labor. Irrigation belongs, in fact, to a highly advanced condition of agriculture, and can only be applied to lands of high value or capacity in the hands of intelligent owners.

But it is clearly manifest at the present time, if it never was before, that the farmer, or other cultivator of the soil, who would succeed in keeping abreast of our progressive age must labor more intelligently, must greatly increase the productive capacity and value of his land, and must employ a larger amount of capital in money, or its equivalent in labor and skill, than he has hitherto done. One of the means placed in his hands, by those circumstances which ever favor the enterprising and industrious man, to employ all these, is to make use of the supply of water, from springs, wells, and streams, which may be available to nourish and increase his crops when rain is withheld, and their growth is consequently arrested.

CHAPTER II.

IMPORTANCE OF AN ADEQUATE SUPPLY OF WATER.

Water is not only necessary for vegetable growth, but it is well established that to a great extent the amount of growth depends upon the quantity of water supplied to a crop. Years ago, when a large portion of the country was covered with forests, and when the cleared soil was well filled with the decaying remains of the removed woods, the produce of the newly cleared fields was more than double that of to day. Then the soil was absorbent of water, it was not subjected to the influence of sweeping winds; the rain-fall was held in the soil for a longer time, and did not pass off in immediate freshets and floods. Consequently the crops had a constant supply of water, and their yield was a maximum one. As a coincidence might be cited the comparatively large average yield of the soil, in the so-called moist climate of England and Ireland. "So-called moist," because, as it happens, the annual rain-fall in our so-called dry climate, is nearly, if not quite, double that of Great Britain. Here the rain-fall is over 40 inches in the year, there it is not much over 20 inches. But the English climate is insular, and is influenced by the moist winds of the ocean, and the fogs from the Gulf Stream. The evaporation from the soil is therefore reduced to a minimum, and the light rain-fall, more constant than with us, and consisting of frequent light showers, is ample for the needs of vegetation. On the contrary our climate is continental and subject to the influence of dry winds, and a higher temperature, and our heavier but more inconstant rainfall is found inadequate. Hence our low average of those crops which need a large quantity of water for their maximum growth, and hence the ineffective efforts of American farmers to reach the high averages of the crops grown in England.

Some very interesting experiments showing this relation between the weight of grain produced and the quantity of water consumed by the plants, whether evaporated through their leaves, or appropriated by their tissues, were made in 1874, at the Agricultural Observatory of Montsouris, France. The grain grown was wheat. Several kinds of soils and fertilizers were used, which gave very varying results, but the variety in the amounts of the product was remarkably illustrative of the facts proved. The means adopted for determining the results were the most complete, and there is no reason to doubt the entire accuracy of the conclusions reached. The results are given in the following table :

I.—*Table showing the total quantity of water evaporated and the grain produced; also the quantity of water consumed for one pound of grain in nine experiments with various fertilizers.*

	Pounds of water evaporated.	*Pounds of grain produced.*	*Pounds of water for one of grain.*
No. 1	1,616	0.6	2,693
" 2	1,512	0.8	1,890
" 3	4,703	2.4	1,960
" 4	2,202	2.7	816
" 5	3,262	2.9	1,125
" 6	4,327	3.1	1,396
" 7	4,751	5.5	864
" 8	7,417	9.2	806
" 9	7,702	10.6	727

The production of straw was very nearly double that of grain in every case, and the increase constant and regular.

In the very exhaustive experiments which have been made by Mr. J. B. Lawes, of Rothamstead, England, to ascertain the amount of water consumed by a growing crop of wheat, it was very clearly shown, that for every pound of dry matter produced, 200 pounds of water was evaporated, and that for every pound of mineral matter assimilated by the crop, 2,000 pounds of water passed through the plant. Mr. Lawes therefore declared, that for a maximum crop of wheat, in England, the supply of rain water was totally inadequate. Leguminous plants, (beans, clover, etc.,) required a still more abundant supply of water than wheat, and of course the more luxuriant the growth, the greater the expenditure of water. Comparing the results of Mr. Lawes investigations with those at Montsouris, a striking equality is found. In the maximum crop there grown, 727 lbs. of water were evaporated for one pound of grain and two of straw, giving 242 pounds of water for one pound of total produce. If, as is probably the case, the weight of the roots was included in Mr. Lawes estimate, as it was not in the other, the approach to equality between the two results would be very close indeed. One therefore corroborates the other.

These results show, in a very remarkable manner, the absolute necessity for an adequate supply of water for the successful prosecution of an advanced agriculture. The plants grown in these experiments were supplied with water *at libitum.* Those which grew luxuriantly under the effect of the most active and valuable manure, viz. a mixture of phosphate of ammonia, nitrate of potash, and chloride of sodium—a very complete fertilizer—are seen to have consumed a very large quantity of water, and nearly five times as much as those which grew most feebly.

The measure of the water consumed may thus be considered as the measure of the capacity of the soil to furnish its product, for it is clear that if this large quantity of water was not supplied, the excessive product of grain could not have been grown. If this conclusion be correct, we have at once a satisfactory explanation of the hitherto strange fact that our best farmers, in no way less skillful or less enterprising, and with no less fertile soil, than the English farmers, can very rarely reach, and still more rarely surpass a crop of 40 bushels of wheat per acre, while in England 64 and 66 bushels are common with the best farmers. Taking the minimum quantity of water, (viz. 727 lbs.) evaporated for a pound of grain, a harvest of 40 bushels of wheat per acre, would consume, or pass through its leaves, an amount equal to 6 inches in depth, over the whole surface of the ground. But this is not a complete statement, for the average result of a large number of experiments made in the previous year, and these results as well, prove that a crop of wheat of 40 bushels per acre, *may* consume, or evaporate, through its leaves, a quantity of water equal to a rainfall of over 17 inches; for the less vigorous the growth, the greater is the proportionate consumption of water, and the yield which consumed 727 lbs. of water for one of grain, was greatly in excess of 40 bushels per acre. If to this consumption of water is added the excessive evaporation from the soil, consequent upon the hot suns and dry winds of our growing season, as well as the loss through the passage of water over the frozen surface of the soil, during our long winters, the totally inadequate supply of water, for a maximum crop, under our now usual conditions, is very evident. It is also evident, that where crops can be grown by irrigation, and an ample supply of water provided, there the success of the farmer will be assured, and there the risks from untimely drouths may be wholly avoided. It is also evident that every where that the conditions permit of it, our grass crops may, by means of irrigation, be made equal to those of the most favored climates, and that the productiveness of our meadows may be increased greatly beyond that which is now possible by the most skillful culture.

But a large portion of our territory is practically rainless and arid. The configuration of the surface is such, that the passage of rain clouds is arrested by high mountains, and the precipitation is confined to very small and elevated areas. This is the case with nearly the whole of our territory west of the 100th meridian of longitude, or a line drawn through the western part of Kansas and Nebraska, from north to south. In this extensive district are found some of the richest soils in the world, which will yield, with irrigation, a yearly average of 30 to 40 bushels of wheat per acre. During the growing season the rain-fall is the least; the greatest amount taking place in the winter months in the form of snow.

The amount of the rain-fall decreases from the 100th meridian, where it is less than 20 inches in the year, to 7 to 15 inches further west, and increases as the Pacific Coast is reached, where it measures $9\frac{1}{4}$ inches in Southern California up to about 23 inches at San Francisco. But the fall is very irregular, depending greatly upon local causes. This is shown by the following facts, derived from scientific observations at various points in California, where the contiguity of the coast range of mountains with that of the Sierra Nevada causes many very surprising differences in the amount of the rain-fall. Thus while at San Francisco the fall averaged 23 inches yearly during 19 years, 14 miles distant at Pillarcito's Dam it averaged during nine years, 58 inches yearly. This irregularity is intensified by the dry winds which absorb moisture to an extraordinary degree; a north wind, hot and dry, which occasionally blows in the San Joaquin and other vallies, has evaporated one inch of water in a day.

The following table gives the range at the various localities for the period mentioned, viz.:

Locality.	*Period.*	*Rain-fall.*
Fort Reading..............	3 years	15.9 to 37.4 inches.
Sacramento................	17 "	11.2 to 27.5 "
Millerton.....	6 "	9.7 to 49.3 "
Stockton..................	3 "	11.6 to 20.3 "
Fort Tejou................	5 "	9.8 to 34.2 "
Monterey..................	5 "	8.2 to 21.6 "
San Diego.................	12 "	6.9 to 13.4 "
Benicia...................	12 "	11.8 to 20.0 "

During these years in which the rain-fall marked the lowest range, the distress amongst farmers was extreme. South of Monterey, in the three years from 1868 to 1871, neither grass nor grain grew. Hundred of farms were abandoned, and stock men drove their cattle, horses, and sheep up into the mountains for food and water. In the Spring of 1870 the great Santa Clara valley was entirely destitute of grass, and the plains of Los Angeles, comprising over a million acres of land, were barren to the borders of the streams. Elsewhere the same effects were visible, and over the entire State hundreds of thousands of horses, cattle, and sheep, starved to death. The estimate of the farmers, in the southern part of the great valley of California is, that but two crops can be secured in five years, without irrigation, but in the extreme south this is to be still further reduced. In 1850 only 7 inches of rain fell at San Francisco.

Further east, in Nevada, Utah and Colorado, where the soil is rich and arable, no dependence can be placed upon the rain-fall, which does not even serve to start the growth of the crops. A great depth of snow, however, falls upon the mountains, which in melting fills the rivers and can be made to furnish an adequate supply during the growing season. Through the whole of this western territory the total supply of water is sufficient to ensure good crops yearly, if it can only be secured and utilized. The first difficulty lies in arresting its escape, and the second in distributing it where it is needed, in an economical manner.

The great valley of California includes an area of 57,200 square miles, which is equal to that of Illinois or Michigan. The area of the lesser valleys is equal to 18,750 square miles, or 12,000,000 acres, susceptible of irrigation. For every one of these acres capable of irrigation, there are three which serve as a water shed, thus, as it were, quadrupling the rain-fall of the valleys, if the water shed of the hills can be utilized.

The area of land that may be brought under irrigation in other parts of the comparatively rainless district, and the area of water shed, has about the same relative proportion, but are of far greater extent. Altogether, the increase of wealth that must accrue from the reclamation of these vast fertile tracts, which want only water to cover them with verdure, is beyond computation. But this increase of wealth, great as it would be, cannot fail to be exceeded by that which would result from the general application of irrigation in those parts of the country where only partial watering is needed; and the prevention of losses by drouth, and the ravages of destructive insects to which moisture is fatal, which every year, in one portion or another of the country, reduce farmers profits, or cause them to disappear entirely. As an example the single case of the grass crop may be considered.

The value of the grass crop of the United States, including hay and the products of pasture, is greater than the combined value of all other crops. This statement will doubtles be a surprise to many, nevertheless it may be substantiated by the following figures.

The total hay crop of 1870 was 27,316,048 tons, the average value of this at a moderate estimate would not be less than $10 per ton, or over 273,000,000 dollars. The total dairy products, which should be credited to pasture, were estimated, in 1870, as 1,000,000,000 lbs. of butter, 100,000,000 lbs. of cheese, and 400,000,000 gallons of milk sold or used. The total value of these is not less than 400,000,000 dollars. Then there should be credited to the grass crop, in large part, the value of the wool and lambs produced, or at least 100,000,000 dollars; also one half at least of the value of the yearly increase of live stock, which is supported on grass the greater part of the year, and this would reach a sum of 200,000,000 dollars. To place these in tabular form would further impress the importance of the grass crop upon the mind of a reader; this may be done as follows:

Yearly value of the hay crop	$273,000,000
" " of dairy products, produced from grass	400,000,000
" " of lambs and wool, due to pasturage	100,000,000
" " of increase of other live stock	200,000,000
Total annual value of the grass crop	$973,000,000

This vast amount is in excess of the value of all the rest of our farm products, in which may be included cotton, corn, wheat, and other grains.

When we consider that by a complete system of irrigating our grass lands alone, the crop could easily be doubled in value, the immense importance of the subject to the agricultural interest of the country is at once seen. There are, comparatively, few cases in which some system of irrigation, more or less complete, could not be applied at least to grass lands, or to now useless lands that could be turned into luxuriant meadows.

But there is still another view of this matter which ought to be considered. It is not only true that water is needed to supply the requirements of plants, but when used in irrigation, it brings within reach of the plants a largely increased amount of nutriment.

Water is the universal solvent. No water in its natural condition is pure. The water of springs and streams holds in solution or suspension a quantity of mineral and gaseous matters, that possess high fertilizing value. The rain water washes the soil, and whether it flows over its surface or percolates through it to the subsoil, it takes up in its course a portion of the soluble matters which it meets. Thus the water of the earth contains lime, magnesia, soda, potash, iron, sulphur, silica, ammonia, carbonic acid, nitric acid and oxygen, in solution. Besides this, many solid substances are held mechanically and in suspension, and are deposited whenever the flow is arrested and the water becomes still.

In Professor Geo. H. Cook's valuable work on the Geology of New Jersey, the following examples are given :

Analysis of water of the Delaware river, made by Henry Wurtz, N. J. State Chemist.

	Grains.
Whole solid matter contained in a gallon	3.97
Consisting of Carbonate of lime	1.30
" " Carbonate of magnesia	0.89
" " Carbonate of potash	0.17
" " Chloride of sodium	0.11
" " Chloride of potassium	0.01
" " Sulphate of lime	0.19
" " Phosphate of lime	0.14
" " Silica	0.50
" " Sesqui-oxide of iron	0.03
" " Organic matter containing ammonia	0.63

The water of the Delaware is considered as exceptionally free from impurities. It is interesting to notice the composition of its impurities in connection with the practically inexhaustible fertility of the flats of this river, which are annually overflowed and thereby enriched.

A comparison of the solid matters contained in 100,000 parts of the waters of several of our rivers is here given, as follows, viz.:

Rivers	*Passaic.*	*Schuylkill.*	*Croton.*	*Hudson.*
Solid contents	12.75	9.41	18.71	18.48
Inorganic	7.85	7.29	11.32	14.52
Organic	4.90	2.12	7.39	3.96

Numerous other examples might be given were they needed ; it will be sufficient for the purpose to notice that these examples are taken from streams, the waters of which were carefully examined, with a view to their value and use for domestic supply of various neighboring cities; and if these waters, selected for their purity, contain so much foreign matter, how much must be contained in those turbid streams, the waters of which are not only highly charged with soluble matter, but carry in suspension solid matter of which vast banks are sometimes deposited in the course of a few weeks or months.

The value of all the water which now passes away uselessly, but which might be arrested and made to deposit on the soil, or convey to the roots of crops, its burden of fertilizing matter, if it were made useful in irrigation, is more than can be readily calculated. An estimate made by Hervé Mangon in his work entitled *Experiences sur l'emploi des eaux dans les irrigations*, of the yearly value of the solid matter conveyed into the ocean by the river Seine, may be cited. He says: "each 200,000 cubic meters of water employed in irrigation, will produce a quantity of alimentary substances equal to one average butcher's beef. Then the waters of the Seine that are lost from the services of irrigation carry into the sea the equivalent of one fat ox every two minutes, or 720 every twenty four hours, or 262,800 in the year." As compared with American rivers the Seine is a small stream; what then might be the value of the Missouri, or the Mississippi, with its affluents, or any one or all of our other rivers and streams, great and small, that now pay no tribute to us in this direction in any way whatever.

CHAPTER III.

THE AMOUNT OF WATER NEEDED FOR IRRIGATION.

There are but few fields or gardens so situated that water may not be applied to them in one or more of the methods which have been at one time or another, or may be, adopted to irrigate the soil. The only prerequisites are, the supply of water and the power to bring it into such a position that it can be spread over the land. Where, however, the cost of procuring and applying water will be greater than the profit to be derived from its use, it may be concluded that there irrigation is impossible. There are some lands situated so far above the supply, that the cost of raising the water and of providing reservoirs to receive and hold it until it could be distributed, would be greater than the value of any benefits likely to accrue from its use. There are others so low that to irrigate them, without at the same time providing for a perfect system of sub-soil drainage, would be to turn them into marshes and ruin them for agricultural purposes. In these cases, if the cost of drainage should exceed the value of the benefits received from the land, it would manifestly be impossible to irrigate them.

On the other hand, where these hindrances do not exist, there are very few physical features of the land that could stand in the way of irrigating it. Level lands, or lands level in one direction with a slope in another; lands sloping in every direction ; hill sides either of moderate slope or such abrupt slope that terraces must be made to retain the soil ; all these may be prepared by simple methods of engineering to receive any supply of water that can be economically brought to them. Equally those lands which happen to lie beneath the level of a stream or tidal river ; a marsh, submerged wholly or partially at certain seasons, or land in similar situations, but not overflowed, may frequently be brought under reclamation and made subject to drainage and irrigation with great profit.

There are also numerous tracts of lands along the borders of many rivers and streams that have been washed and injured by freshets so as to be in their present condition worthless for cultivation, which at a small outlay may be covered with new soil of a most fertile character, and again rendered useful and profitable by the use of appropriate methods of irrigation. Besides these, there are extensive tracts of land at the mouths of tidal streams or estuaries, or at the confluences of large rivers, which are always under water or exist as mud banks, which may be reclaimed by judicious engineering, and converted in a few years into agricultural land of the richest quality. All these processes belong to the art of irrigation, and the cases in which one or another of them are impossible of application are very rare indeed.

The supply of water is a more serious consideration than the shape or configuration of the land. Where this is not naturally available no art of the engineer can provide it. The only safe dependence is upon streams or springs, and reservoirs in which the rain-fall of winter and spring may be gathered and stored. Wells can only be depended upon for such a small supply as would serve to irrigate a garden or small market farm, where the large value of the crops would admit of the cost of raising water for a lengthened season and storing it in reservoirs for use in emergencies. The idea that artesian wells may be made a source of supply for completely irrigating large tracts of land, if ever held by any over-sanguine persons, must be abandoned. For partial irrigation they may be made available, but the quantity of water needed for the irrigation of a few acres of land only, in localities where there is no summer rain-fall, as upon our Western plains, is far beyond the capacity of any artesian well to supply, unless it be one of extraordinary volume.

It is very important that the quantity of water needed for irrigation should be accurately estimated. A mistake in an estimate may lead to the construction of inadequate works, and the useless expenditure of much money. Estimates generally err upon the side of insufficiency rather than otherwise, and much error has been spread abroad by persons and journals having considerable influence. Not long ago the "Scientific American" editorially announced that one artesian well would supply a farm of 640 acres upon the plains with water for irrigation, and would also form a nucleus for many large stock farms." The late Horace Greeley, who, although an enthusiast upon this subject, was more nearly correct, thought one artesian well would serve to irrigate a quarter section of land, or 160 acres. The wildly excessive estimates of the value of a certain amount of water might be easily disproved by the careful use of a few figures, and a little common sense. For instance, let any person who has ever watered a garden plot and who knows the effect of one inch in depth of water upon a dry soil, consider the following facts, and then apply them to the matter in question, and he will readily see the absurdity of the estimates above referred to.

1st. There are 6,272,640 square inches in an acre.

2d. One inch of water, or a stream one inch wide and deep, flowing 4 miles an hour, will equal 6,082,560 inches in 24 hours.

3rd. Therefore 1 inch of water flowing 4 miles an hour, for 24 hours, will cover one acre nearly an inch deep.

4th. One inch of water per week equals 52 inches per year, or more than the yearly rain-fall.

5th. Therefore 1 inch of water should serve to irrigate only 7 acres once a week, at least as well as the average rain-fall.

6th. One inch of water flowing 4 miles per hour, equal one and one-fifth quart per second.

7th. One quart per second, flowing for 24 hours, will cover an acre five-sixths of an inch deep.

8th. One inch of water flowing 4 miles an hour is equal to 18 gallons per minute, or 1,080 gallons per hour.

9th. An artesian well, 6 inches in diameter, would give a stream of 28 square inches, and would deliver 32 quarts per second, if the flow were at the rate of 4 miles an hour.

10th. Such a well would furnish an inch of water per day for 28 acres, or an inch a week for 196 acres, which would be a very insufficient quantity to irrigate dry open soils in places where the climate is arid.

11th. The cost of such a well would be at least $5,000 to $10,000, or more than the value of the land when irrigated.

The estimates made by various authorities upon irrigation, as to the quantity of water needed, vary considerably. As a rule, the quantity of water used by some irrigators, would seem to be extravagant. Thus we find standard authorities upon irrigation, and practical irrigators, recommending and using quantities of water varying from one to four quarts per second, continuously flowing for 24 hours for each acre, at intervals of from five to fourteen days. It is evident, however, that the quantity of water needed to moisten the soil thoroughly, depends on certain conditions, which are very variable.

These conditions are :

First—the nature of the soil.

Second—the character of the climate.

Third—the nature of the subsoil.

As to the Soil.—Soils differ greatly in their power to absorb and retain water. Those which absorb most water retain it for the longest time. The power of absorption is due to the surface attraction of the particles of soil for water. The finer the particles of the soil, the greater will be the amount of water absorbed, because the total surface of the particles is greater, and the longer will it be retained. Thus a soil consisting of coarse gravel will not retain water. A soil of pure quartz sand will absorb but a small quantity, and will soon part with it, while a fine alluvial soil will absorb a large amount, and retain it a long time. The following table gives the results of experiments made by Schübler, to determine the capacities of different soils for water and their comparative power of retaining it. In these experiments the different soils were thoroughly wetted with water up to the point of saturation, and the increase of weight noted; this is shown in the first column. In the second column are given the quantities of water which evaporated in four hours, the samples of soil being spread over equal surfaces.

	Per cent of water absorbed.	*Per cent of water evaporated in 4 hours.*
Quartz sand	25	88.4
Limestone sand	29	75.9
Clay soil (40 per cent sand)	40	52.0
Loam	51	45.7
Common arable land	52	32.0
Heavy clay (20 per cent sand)	61	34.6
Fine Carbonate of lime	85	28.0
Garden soil	89	24.3
Humus (peat or decayed vegetable matter)	181	25.5

Thus the greater capacity a soil possesses for the absorption of water, the longer it retains it. It is obvious that upon this depends to a very great extent the quantity of water that will be needed for the irrigation of any particular soil. Before any calculation, as to the needed supply, can be made, this point will have to be duly considered and determined by the irrigator or hydraulic engineer. The difference arising from the variations in the texture and composition of soils has been closely studied by the French irrigators and engineers. M. Gasparin, who stands at the head of the numerous writers upon this subject in that country, states that a soil which contains 20 per cent of sand needs to be irrigated but once in fifteen days, while under similar circumstances, another soil which contains 80 per cent of sand, should be irrigated once in five days. The difference would be still greater between soils varying still more in their character, and less with those which may be classed between these limits.

As to the Climate.—As already stated, by far the largest portion of the water which falls upon the earth's surface is removed by evaporation. Observations made at Abbot's Hill, England, by Mr. Dickinson, during eight years, showed that 90 per cent of the water which fell in the summer, or between April 1st and October 1st, was removed by evaporation, and only 10 per cent found its way into the drains which were from 3 to 4 feet deep. The total quantity of water which fell in those six months was equal to 2,900,000 lbs. per acre, and of this more than 2,600,000 evaporated. It should be remembered that this occurred in a moist, cool climate, the verdure of the meadows in which is hardly equalled in any other country, unless it be in the still more humid Ireland, "the emerald isle." In England showers occur almost daily, and the winds blowing in any direction from the sea, seldom more than a hundred miles distant, and generally much less than that, are charged with moisture; the maximum summer temperature rarely reaches 80 degrees, and also from the more northern latitude, the sun's rays fall at a comparatively low angle; if then, under these conditions, evaporation carries off nine-tenths of the moisture from the soil, what allowance must be made in our climate, where the atmosphere is drier, the summer temperature 20 degrees higher, and where the sun's rays fall upon the surface more directly, and more ardently. And further, if a large allowance must be made in those parts of the country where the rain-fall amounts to 40 inches and over, how much more liberal must the allowance be for those districts where the rainfall is 10, 15, or 20 inches, and where the winds, almost completely deprived of moisture, thirst intensely for it? Here is a consideration of great importance, and one which cannot be disregarded.

It will be evident to the thoughtful reader, that much will depend upon the condition of the surface of the soil maintained by the cultivator. The amount of evaporation can be largely controlled by keeping the soil in a finely divided and mellow condition, in which it holds its moisture with the greatest tenacity. But there are crops, such as wheat, oats, etc., which do not admit of cultivation during their season of growth, and these must necessarily require a larger quantity of water than such crops as corn, or roots, which can be cultivated.

In the dry and hot climate of Provence, a district in the south of France where irrigation is extensively practiced, it has been found necessary to use for each watering of the soil a volume of water equal to a depth of 3½ to 4 inches over the whole surface every 10 to 12 days,—the usual interval between the waterings. This is equal to about 24 cubic inches, or nearly half a quart per second, continually flowing for each acre of surface. This allowance, which in French measures is equal to 1 litre per hectare, or 61 cubic inches (=1 litre) per 107,640 square feet (=1 hectare), is the basis for all contracts between the government which controls or supervises the water supply, and the owners of the canals (*compagnies concessionaires de canaux*), and between the latter and the farmers who buy the water from them. It is the official and legal unit of supply, as it were, and is a valuable general indication, applicable to any locality or country, where water may be used to irrigate soils of different characters and for different crops. This may be taken as the mean quantity, to be decreased or enlarged as circumstances may necessitate the change.

As examples of the nature of these varying circumstances, the following are cited: M. Hervé Magnon (a frequently quoted authority in works upon irrigation, and already referred to here), determines the limits of supply as from one to four litres per second per hectare, which is equal to from one pint to two quarts per acre, per second, continuously flowing. Gardens and market gardens require the larger extreme. M. Pareto, another French author, in his work upon the irrigation and drainage of lands, (*Irrigation et assainissement des terres*), mentions some cases in which a quantity equal to one quart per second was sufficient to effectively irrigate eight acres. This may be taken as the extreme minimum limit of supply, very rarely occurring, and altogether exceptional.

The Italian canals, which irrigate 1,600,000 acres, supply 24,000 cubic feet per second for this area. This is equal to one cubic foot (30 quarts) for 66 acres, or a flow of 26 cubic inches per second per acre; or very nearly one quart (which is $57\frac{3}{4}$ cubic inches) for each two acres. In that country the rain-fall equals 37 to 38 inches per annum, the most of which occurs in the irrigating season, when there are on the average 71 rainy days in the six months from March to October. There the summer temperature is from 70 to 90 degrees. It will be observed that the climatic conditions of Italy closely approach to those of the rainy portion of the United States. The mean water-supply may therefore be taken as closely approximating the quantity required in this country—viz., one pint per second, constantly flowing, or 10,800 gallons, or $172\frac{2}{10}$ cubic feet every 24 hours for every acre. In India, one cubic foot per second is made to serve for 200 acres of grain crops. In some parts of Spain the same quantity serves for 240 acres; in others the same quantity is spread over 1,000 acres, and the legal allowance in some recent Spanish grants, varies from 70 to 260 acres per cubic foot per second. Rice culture requires a supply equal to one cubic foot per second, for each 30 to 80 acres. These examples will serve as a basis for calculations, needed to meet the widely different circumstances which exist in the United States, where possibly the variations of soil and climate, over our extensive territory, are unparalleled in any other single country in the world.

As to the Subsoil.—This point needs very little elucidation. From the preceding remarks, the effects of loose or compact subsoils will be seen to exert a considerable influence upon the requisite water supply. There are soils which rest upon open, coarse, gravelly subsoils, which may be compared to a sieve. Other soils, with retentive clay subsoils, furnish examples exactly the reverse. These are unusual cases, but as they may occur, they ought to be considered. It does not seem necessary to discuss this point further than to call attention to the importance of ascertaining the character of the subsoil of any tract which it is proposed to irrigate, as a serious element in calculating the needed supply of water. The minimum direct loss through absorption by the subsoil should not be estimated at less than 15 per cent of the supply, and a much larger allowance should be made when, upon examination, the subsoil is found to consist of coarse sand or gravel.

When we consider the quantity of water needed for irrigation, it is clearly seen that springs are rarely of sufficient volume to be of material value, excepting for meadows, and then only for small surfaces or partial watering. They may be made, however, to serve an important purpose in such cases as these where the area to be irrigated is small. Storage reservoirs, in which is collected the water of those temporary courses, which flow only when the rain-fall is largest and the amount of evaporation is least, may be made important sources of supply. It is by means of these that a large portion of the irrigation needed to make the dry plains of India fruitful is accomplished. But by far the most important sources of water for irrigation are rivers and streams. In these there is an abundant supply, and there is generally ample provision for elevating the water, by means of dams with canals or by water wheels, to the highest portion of the adjacent land which is to be irrigated.

The scope for the utilization of rivers and small streams in irrigation in the United States is of vast extent, and the statement which has been made that there are 500,000 homesteads in the country that could be brought under a partial or complete system of irrigation, does certainly not overestimate the reality, but on the contrary is doubtless greatly below it. It is for every cultivator of the soil to closely scan his own resources in this respect, wisely determining to turn them to accouut as soon as he shall have discovered their existence and perceived how to employ them. The cost of works for irrigation will be greatest where the area to be irrigated is the smallest, as for instance in gardens and market gardens; it will be least in the case of meadows, where the distributing canals are permanent in character, and between these extremes upon arable lands, where for each crop the surface must be disturbed, and furrows for spreading the water must be made anew at each plowing. The cost will also vary greatly, as the facilities for procuring and elevating the water may differ. But it may be accepted as beyond doubt that there are few gardens, market farms, orchards, or meadows that might not be brought under a more or less systematic irrigation, and few localities near the borders of rivers in the great Western plains, or other rainless localities, in which the present arid desert may not be redeemed and made to blossom and become fruitful beneath the beneficient influence of the fertilizing waters which now flow uselessly by them.

CHAPTER IV.

IRRIGATION OF GARDENS.—THE SUPPLY OF WATER.

Gardens and market farms, by reason of the greater value of the crops raised upon them, in constant succession, will permit the application of more costly methods of irrigation than any other cultivated grounds, and from their smaller area there is less difficulty in procuring an ample supply of water. Few gardens are so situated that water can be procured from a stream without the employment of a water wheel or other motive power, a force pump, and pipes laid underground, and a reservoir in which water may be stored when not needed. But nearly every one may be supplied from a well by the use of a windmill. A windmill of the smallest size made, and of the best construction and self-regulating, costing about $100, is able to raise two quarts of water per second to a hight of 25 feet. A windmill may be constructed by any fair mechanic at a cost of from $10 to $25, which will answer every purpose of those manufactured and sold at higher prices, excepting that of regulating themselves to the varying forces of the winds. A mill of this character may be fixed in a frame over the well, and the arms, of which there may be six, eight, or more, with fans fixed so as to present their faces at an angle of 45 degrees to the wind, are kept in position by means of a vertical vane behind them. Another, which consists of six arms mounted upon a rotating frame, carries cloth sails. This mill requires to be changed as the wind changes, and a ladder is attached to the frame upon which it is mounted for this purpose. The frame on which to mount it may be of timber, as shown in the engraving (fig. 1), or it may be a stone or brick building if desired for a substantial machine for heavier work. The power is constructed in the shape of arms—shorter or longer, according to the power needed—fixed to a center-wheel or hub, which is mounted and keyed on to an axle. Sails are carried on these arms, of sail-cloth or heavy sheeting, of a triangular shape, as shown in the engraving, which are fastened closely to one arm and by a cord in the corner, shown at *a*, a foot or less in length, to another. This gives sufficient inclination backward to the sails to gain the motion required with a front wind. On the axle is a crank-wheel, *b*, which moves the rod to be connected with the pump, or it may be connected by means of pulleys and bands to get an upright rotary motion, or a pair of bevel-wheels will give a horizontal rotary movement. A frame, *c*, is carried on a circular table, on which it may be revolved so as to enable the sails to be presented fairly to the breeze ; a box, *d*, at the rear end of the frame is weighted with stone, to balance the weight of the arms and sails. A pin passed through holes in the circular table retains the frame to the position needed, and keeps the sails faced to the wind.

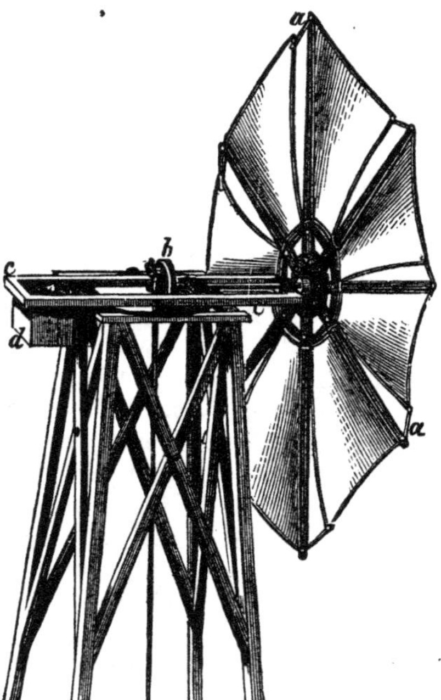

Fig. 1.—WINDMILL WITH SAILS.

A mill with arms six feet long may be made to do work equal to one-fourth of a horse-power, if all the working parts are well fitted and kept well lubricated, as all machinery should be. When out of use, the sails are untied and removed, or they may be furled and clewed to the arms until again required.

A one-horse railroad-power would also serve a useful purpose in raising water from wells into an elevated reservoir, where it could be stored for use. For small

Fig. 2.—SQUARE TANK.

gardens the water from the roofs of buildings may be collected in tanks or cisterns raised a few feet above the level of the ground.

A round tank, hooped with iron bands, 12 feet deep and 15 feet in diameter, will hold over 15,000 gallons. A square tank (fig. 2) may be made of jointed and matched planks, which are forced closely together by wedges, acting upon a timber frame which encloses the planks. This is the cheapest kind of tank that can be made. One 16 feet square and 10 feet deep will contain nearly 20,000 gallons. Tanks of this character can only serve for small gardens, or to store water which is pumped at night for use during the day time. Either of these tanks, if filled during the night (to do which will require a stream from a pipe of an inch and a half in diameter constantly running), and replenished during the day, will furnish enough water to give more than one inch in depth over an acre of surface. This is the least quantity that could be depended upon in a dry season for any effective purpose, and would need repeating after an interval of four to seven days, so that the maximum effort of a tank of this size, with a well, windmill or horse-power attached, would suffice only in an emergency to water four to seven acres of land. Where the ground to be irrigated is of larger extent, the tank room and water supply must be enlarged, or the diameter of the pipe and power increased. The capacity of the pipe increases as the square of the diameter, by which is meant that if the diameter is doubled the capacity is quadrupled. Thus if a pipe one inch in diameter supplies one quart per second, a pipe of two inches diameter will furnish four quarts per second (or two multiplied by two), and a pipe three inches diameter will yield nine quarts (or three multiplied by three), per second. At the same time the power must be increased in proportion to the amount of water elevated, or disappointment will result. In estimating power a large allowance must be made for loss. A horse working in a railway-power can only raise an equivalent of three-fourths of his weight; the rest disappears in friction; and when a stream of water is forced through a pipe of small diameter for a considerable distance, the loss of power in friction is very large, and another fourth of the horse's effort must generally be allowed to compensate for it. One horse may be expected to raise 180 quarts one foot high every second, or 6 quarts to a hight of 30 feet. The small size windmills are about one-sixth of one horse-power.

Where streams are available, the supply of water will be found most ample and most economical. No storage tanks are needed in which the water must remain for a time, that its temperature may be raised nearly to that of the soil, as when wells are used. The water may be taken directly from the stream and flowed upon the ground. A low dam of two feet in hight may be constructed of planks across the stream, by which power to run a small undershot wheel may be secured. Where there is facility for backing the water to a greater extent, or of procuring a greater fall, a breast-wheel may be used. A dam four or five feet high will be sufficient for a wheel of this kind, if the stream is four feet wide and six inches deep, and runs with a velocity of two miles per hour. Such a stream with this fall of water would give sufficient power to elevate 11 quarts of water per second a hight of 30 feet, or a sufficient supply for about 12 acres of ground, or more in proportion to the less hight that the water would have to be raised. To calculate the nominal horse-power furnished by a fall of water, the velocity of the stream in feet per minute, the hight of fall, and the sectional area (the width and depth) of the stream in square feet, must be multiplied together, and by $62^1/_2$, and divided by 33,000.

For instance, if the stream is found to be 5 feet wide at the surface, and 3 feet at the bottom, with banks evenly sloping from the surface to the bottom, the mean diameter is found by adding the surface and bottom widths together and taking half the sum. In this case the mean width will be 4 feet. If the depth in the middle is 6 inches or half a foot, this mean width is multiplied by the depth and the product is the sectional area, which in this case is two square feet. To find the velocity of the stream a thin shaving or other light floating substance is thrown upon the surface, and the exact time in which it moves over a definite distance, say 10 rods or 165 feet, is carefully noted by walking along the bank watch in hand. Let this time be supposed to be one minute. Then the sectional area of the stream being 2 feet, this is multiplied by 165 and the product 330 is the number of cubic feet of water passing down the stream in one minute. A cubic foot of water weighs $62\frac{1}{2}$ pounds, therefore 330 cubic feet weighs 20,625 lbs. If the dam is 4 feet high we have 20,625 lbs. of water per minute falling 4 feet, which is equal to 82,500 lbs. per minute falling one foot. This would, as a matter of course, exactly balance the same weight rising the same hight. The whole power of a horse attached to suitable machinery is equal to that necessary to raise 33,000 pounds one foot high in a minute. The force exerted by the falling of 82,500 pounds in a minute is equal to $2\frac{1}{2}$ horse-power. But a considerable allowance must be made for friction, when waterwheels are used, and especially where the fall is so small as here supposed. It would not be safe to expect to gain more than one half of the whole effect in this case. The power gained would therefore, under ordinary circumstances, be about $1\frac{1}{4}$ horse-power, or sufficient to raise about 40,000 lbs. or 20,000 quarts a foot high per minute. This is equal to about 11 quarts, 30 feet high, per second.

If it is found necessary to store the water thus elevated so as to extend the area that may be irrigated, cisterns of substantial construction will be required. These should be of brick or stone laid in cement, or hydraulic lime, and strengthened with buttresses upon the outside. A bank of earth should then be heaped up around it and sodded, and if the bank be terraced, it may be utilized by planting it. A remarkably elegant structure of this kind is to be seen in a market garden at Astoria, Long Island. It consists of a large cistern of stone work surrounded by earth sodded in part and in part planted, and surmounted by a rustic stage and summer-house built of cedar boughs and roots. Above the whole, towers a powerful windmill which serves to pump the water from a well near by into the tank and force it from thence into the extensive greenhouses and other buildings upon the farm. Although the cost of such a structure is large, yet it is in such a case as this no more than a necessary outlay of capital, without which the business could not be carried on, and is simply an expenditure made in a true spirit of economy.

Such a tank of considerable size and great utility (see fig. 3), may be dug in the ground at the highest part of

Fig. 3.—BRICK CISTERN.

the garden, to such a depth that the soil excavated will make a retaining bank to support the portion of the wall that is above the surface of the ground. This tank, which is circular, may be covered with an arch of brick work, and may be surmounted by a tool-house or other useful building. In this case a brick shaft 2½ feet thick each way should be build in the center from which the arch would spring to the circular wall of the cistern ; the wall should be 9 inches thick and the bottom may be either of bricks laid flat or of cement laid upon the earth.

This cistern, if 20 feet in diameter and 12 feet deep, would hold 30,000 gallons, or enough to water over three acres at one time. If the cistern is open the wall could slope outward, making an inverted frustrum of a cone (as seen in fig. 4), 32 feet wide at the surface and 8 feet wide at the bottom. The earth thrown out at the bottom will form a support for the upper portion of the wall. But before the wall is built the earth thrown out should be solidly rammed down in layers made hollow or

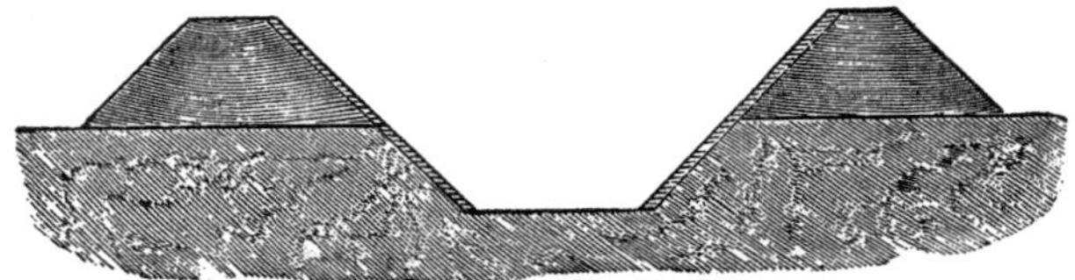

Fig. 4.—OPEN CISTERN.

of the form of a basin. The form is shown by the curved lines in that part of the engraving.

There is a large variety of pumps adapted to the purpose of irrigation, but the severe uses to which they are put make it desirable to have only those which are constructed entirely of metal or wood. Leather valves are soon worn and become useless, causing delays, and serious loss of time in repairs. The double action force pumps, with metal valves, or the rotary pumps of the ordinary kinds with metal pinions which work into each other similarly to cogwheels, or those which work upon the old-fashioned principle of the Archimedean screw, but which nevertheless are protected by a modern patent are all suitable for this work on account of their durability. A double-acting force pump of the most simple character (fig. 5), made almost entirely of wood, is one of the best for this purpose on account of its cheapness and the ease with which it is kept in working order. It is formed of a block of wood, *A*, *A*, in which two parallel holes are bored lengthwise. In these holes the plungers, *B*, *B*, made of wood—maple being preferable—are worked by rods affixed to a rocking shaft in connection with the power above the ground. Between these holes a smaller hole, shown by the dotted lines, is bored. This bore is made to communicate with the other two by a hole bored from the outside (seen at *C*, that portion shaded and where the letter *C* is seen being afterwards plugged up). A leather valve is placed so as to close the ports of this last hole and turn the current of water into the pump tube. This valve is inserted into a dove-tail mortise, cut in the bottom of the block. A slotted plug, *D*, holds the valve, and is placed and fixed in a proper position in the mortise. The lower portion of the mortise is closed with a plug. To insert the slotted plug a hole is bored and the bottom of the block is sawed into to give room to chisel away the space in which the valve works back and forth. The pump tube may be a log bored and inserted into the block, as shown at *E*. Half-inch iron rods may be used to work the plungers. This simple and useful pump requires for its construction only those materials that are available everywhere, and only such skill as

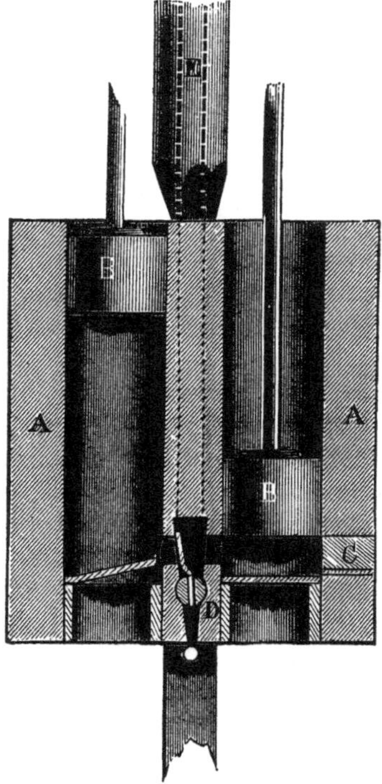

Fig. 5.—DOUBLE-ACTING FORCE PUMP.

is possessed by any village carpenter or mechanic. It has been patented by Mr. Ed. Buzby, of Shamong, N. J. Where metal pumps are preferred, the American Submerged pump made by the Bridgeport (Ct.) Manufacturing Co., and which are entirely of metal and almost indestructible, would be found very suitable. For lifting larger quantities of water a great variety of wholly metallic pumps are manufactured by the Hydraulic and Drainage Company of Brooklyn, N. Y.

CHAPTER V.

PREPARATION OF THE SURFACE.

An adequate supply of water having been obtained, the preparation of the surface of the ground to be irrigated is the next work. For gardens this should be very complete, as the work will be permanent, and the first outlay will be the last, if the work is properly done. The method of laying out the ground will depend greatly upon the nature of the surface. If it is perfectly level, with no perceptible slope in either direction, the method of bedding should be employed. This is done by plowing the land in ridges of such a width as are most convenient for the culture carried on. For market gardens, where horse cultivation is practiced, these beds may be from 20 to 30 feet in width. In smaller gardens, in which the hoe is used and hand labor employed in cultivation, ridges of 10 to 12 feet in width will be found more convenient. Where the spade is used altogether and horses are never admitted, the ridges may be made of even less width ; the dimensions depending altogether upon the convenience or the necessity of the cultivator. The system described applies to each of these cases. The ground is laid out into plots of a convenient size, which run completely across the garden or inclosure, in a direction parallel with that of the main water-furrow from which the supply is to be derived. In case the garden consists of four, eight, or ten acres, or less or more, a proper width of these plots would be 210 feet. This size would be the more convenient, as 210 feet is as nearly as can be had in practice the length of the side of a square acre. Besides, this distance is as great as water can be made to run in a furrow in ordinary garden soil without being all absorbed before it reaches the extremity. Between the plots sufficient spaces will be left for roads, if any are needed, for carts or wagons to go through. These plots are then divided into other plots of the width designed for the ridges. They are then plowed, and the ridges "twice gathered"—to use a plowman's parlance—which means

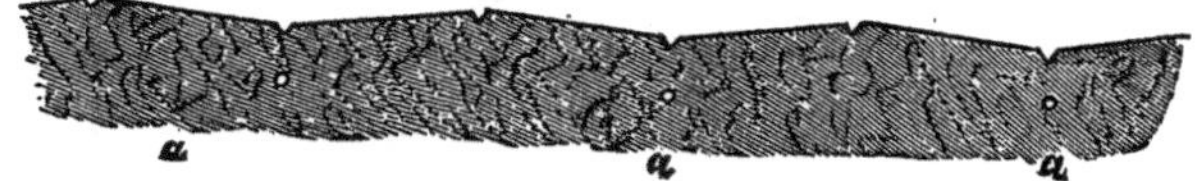

Fig. 6.—OUTLINE OF THE BED.

that a back furrow is made in the center of each of these secondary plots, and the furrows are thrown each way toward the back furrow until the ridge is completed. The ground should then be rolled. Then another back furrow is made over the first, and the ridge is plowed as before, making each of the furrows shallower than the preceding one, so as to leave a gentle slope from the crown of the ridge toward the open furrow on each side of it. The ridges will then show an outline as seen in fig. 6. At the head of each row of ridges or beds the ground is plowed into a headland or ridge, which is thrown toward the first made ridges, and which slopes gradually away from them to the fence or outer boundary of the inclosure, the last furrow made, next the fence, being plowed deeply so as to provide a ditch for draining the headland. The principal canal of supply for the range of ridges below it will run along the crest of this headland, and a canal of distribution will run along the crest of each of the secondary ridges. Each headland or principal ridge, with its canal, and the range of ridges starting at right angles from it, each one of them having its distributing canal, will then form a system of irrigation independent of the other series of ridges. Every seven of these secondary ridges, if they are 30 feet wide and 210 feet long, will occupy one acre of ground. At the foot of each series of ridges will be needed a draining

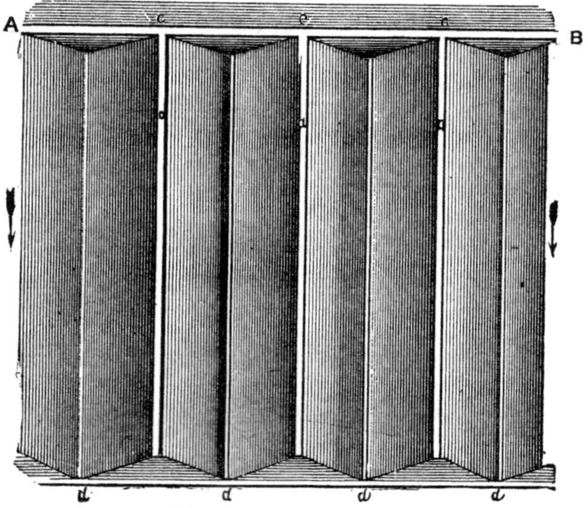

Fig. 7.—SYSTEM OF BEDS.

furrow, unless the ground is underdrained with tile, to carry off the surplus water. A tile drain between each pair of beds or secondary ridges would be the best method of drainage, and the supply of water should be regulated so that the whole is absorbed and none is allowed to flow away unused. The tile drains are shown at *a, a, a,* fig. 6.

The series of beds and canals will then appear as shown

in fig. 7, in which three secondary ridges, *a, a, a,* with the head-ridge, *A, B,* and the canals, *c, c, c,* belonging to each are shown, with an open drain, *d, d, d.* The arrows show the direction in which the water flows. Fig. 8 shows the profile of the ridge and section of the head-ridge with its canal of supply as if they were cut down

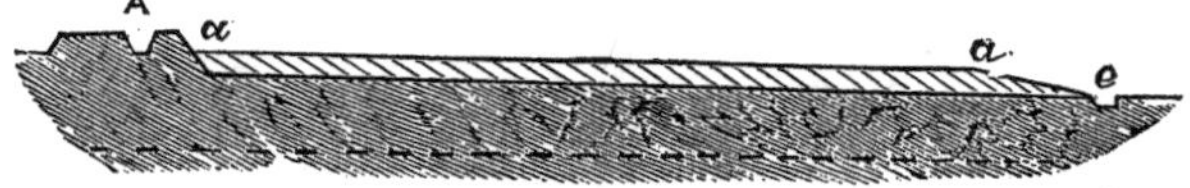

Fig. 8.—PROFILE OF BED.

through the center, *A,* being the head-ridge with its canal, *a, a,* the bed or secondary ridge, *c,* the drain at the foot of the bed, and the dotted line shows the course the tile drain would take below the surface, should one be laid.

Where the ground has a slope in either direction the system to be adopted will be much simpler than the preceding one. At the head of the slope will be placed the canal of supply. This will be the only permanent work undertaken. The method of cultivation of the field or garden will control the method of distributing the water. It will be necessary, however, to cultivate the ground in drills or hills or subordinate beds, upon which the water

Fig. 9.—FURROWS FOR A REGULAR SLOPE.

may be turned when it is needed, leading it by small furrows or canals made with the hoe or a small hand plow in whatever direction, down or across the slope, as may be desired. Generally the arrangement of the canals of supply will be as shown in fig. 9, in which the supply canal is seen at *a,* and the drain which carries off the surplus water is seen at the foot of the slope at *b.* A low ridge separates the latter from the next supply canal. In this method of irrigation the water may be supplied as a thin sheet flowing over a smoothed surface, or as a number of small streams, flowing in a network of courses over the surface, or in regular channels between the drills or rows of plants. The ground may be laid out upon various plans, as the method of cultivation adopted may require. A plan (see fig. 10), adapted for a crop

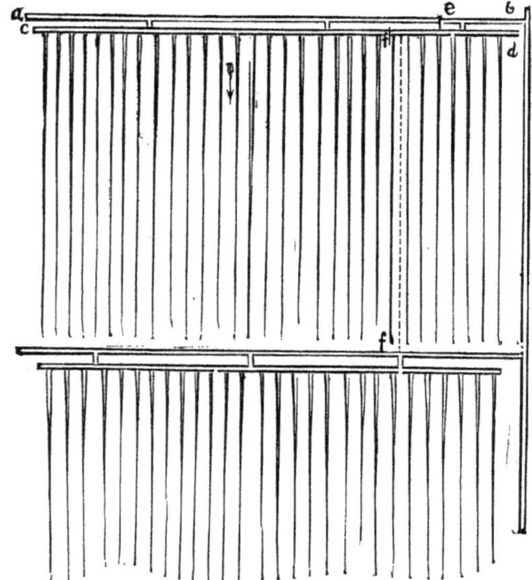

Fig. 10.—ARRANGEMENT FOR HILLS OR DRILLS.

cultivated in hills or drills, each drill forming its own furrow of distribution in which the water may flow, is as follows: A supply canal, seen at *a, b,* is made at the highest part of the ground, with several short canals connecting it with a distributing canal, *c, d.* From this distributing canal the water flows into the furrows, shown by the fine lines. The field is watered in sections by closing the canal at any desired place, as at *e, f,* with a sheet-iron plate or wooden gate, shown at fig. 11, in which is seen the gate, and at fig. 12 the method of its use. Obviously by shutting the canal in this manner the irrigation is confined to the portion of the field circumscribed by the closed furrow, shown by the dark dotted line, *f*, *f*, in fig. 10. The direction of the water is shown by that of the arrow. Where the slope of the ground is too abrupt to admit of very long furrows, a different plan, shown in fig. 13, may be adopted. In this the supply canal, seen at *a*, *b*, is the same as previously described. From this the lateral canals, *c*, *c*, *c*, are made, each of which supplies its own dependent furrows, and no more water is admitted to these canals than will water the surface to which it is tributary. These canals gradually decrease in size until they disappear at the

Fig. 11.—HAND-GATE.

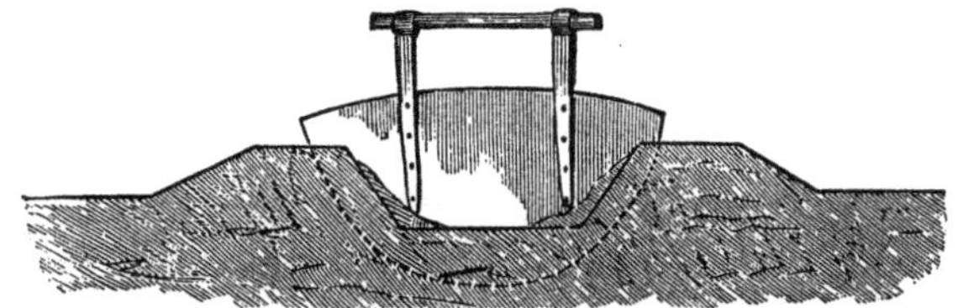

Fig. 12.—MODE OF USING HAND-GATE.

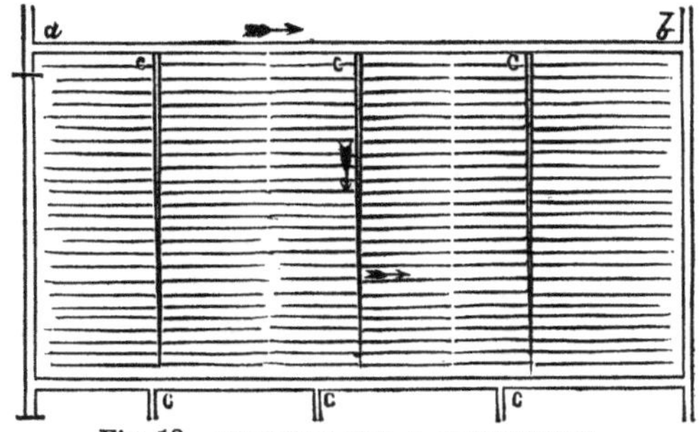

Fig. 13.—FURROWS FOR A STEEP SLOPE.

boundary of the field or garden. The water flowing from these lateral canals takes the direction shown by the arrow. A more elaborate arrangement will be suitable to market-farms, where a variety of crops, each needing especial treatment, are grown. (Such a one is shown in

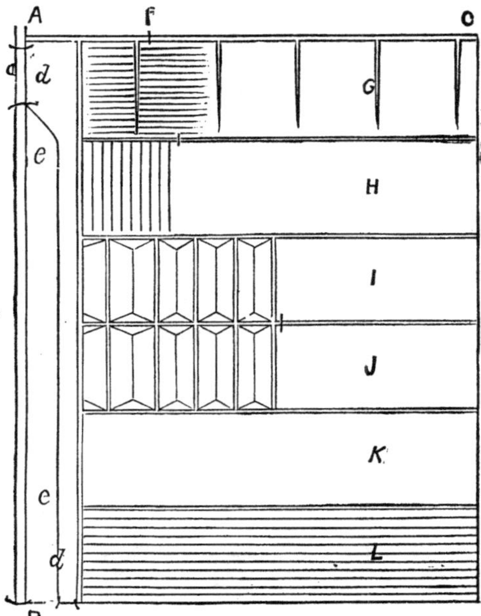

Fig. 14.—ARRANGEMENT FOR A MARKET FARM.

fig. 14.) In this the water is supplied by one or two canals, *A, B* and *A, C,* as may be consistent with the slope of the ground. A road, *d, d,* is laid out at one side of the plot, crossing the supply canal or feeder by a culvert at *a,* a portion of the ground, *e, e,* being retained for cultivation, leaving room to turn a cart or wagon at each end of it. The water is turned from the main supply canal, *A, B,* into the main distributing canal *A, C.* The ground to be cultivated is laid off into plots such as are suitable to the system of culture as at *G, H, I, J, K, L.* These may be irrigated in diverse ways, as for example by long furrows, at *L,* in smaller beds with shorter furrows, as shown at *G,* or in furrows running in an opposite direction, at *H.* The flow of water in the distributing canals is controlled and diverted by means of the hand-gates already described, as at *f.*

A modification of this plan of arranging an irrigated garden is as follows (fig. 15). An alley-way or cart-road

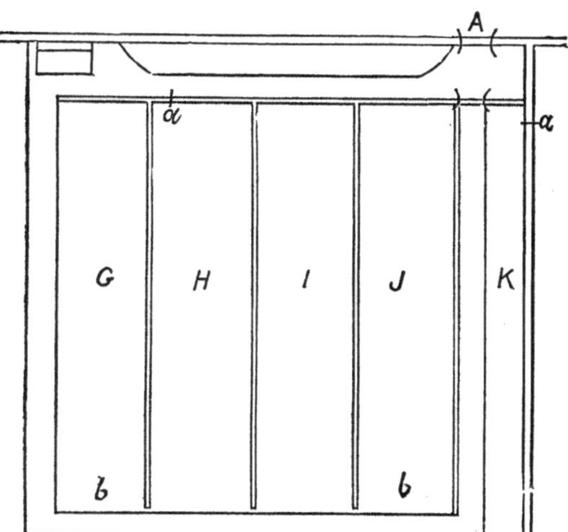

Fig. 15.—METHOD FOR AN IRRIGATED GARDEN.

may be made opposite the entrance *A,* which crosses the canal by a culvert, and a path is continued quite around the enclosure. The beds *G, H, I, J, K,* are watered from the distributing canals as in fig 14, and the flow is diverted and controlled by means of the hand-gates already described, (fig. 12), which when placed as seen at *a, a,* turn the water on to the bed *H.* This water may be directed amongst the hills or drills, wherever it may be required, by means of small canals made with a hoe and the surplus will be caught in the foot drain, *b*, *b*. A great variety of methods may be used with this and the previous plans, so as to meet the necessities of all sorts of crops. The renowned Erfurt cauliflowers are grown in gardens irrigated on such a plan as this; the water flowing in permanent ditches being dipped up with long-handled scoops, and scattered about the plants daily. These cauliflowers are grown upon what was originally low, wet soil, and the ditches serve at the same time for drainage and irrigation.

A plan for a garden very completely irrigated by means of a well or reservoir may be laid out as follows, (see fig. 16). A road passes through the center and around the

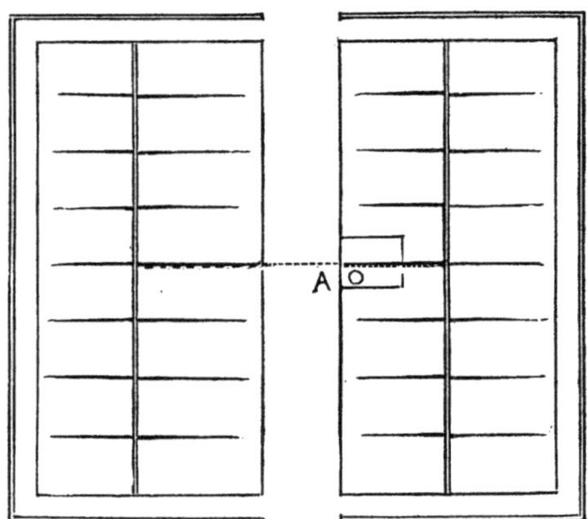

Fig. 16.—PLAN FOR IRRIGATING FROM A WELL.

plot. The well and reservoir, windmill or horse-power, are situated at the highest part of the ground, (see *A*). From this the water is conveyed by channels, (shown by dark lines), to the lower parts of the garden. From these channels it is distributed in small furrows to every row of plants or vegetables. For a small garden this system is doubtless the most perfect of all methods of irrigation by surface channels and furrows ; while for larger ones or market farms, in which the supply can be procured from wells or carried into reservoirs for final distribution, it is equally perfect. The form of the channel deserves consideration. The typical canal or furrow, (shown at fig. 17), is one in which the earth thrown out forms a bank above the channel, preventing the influx of water from

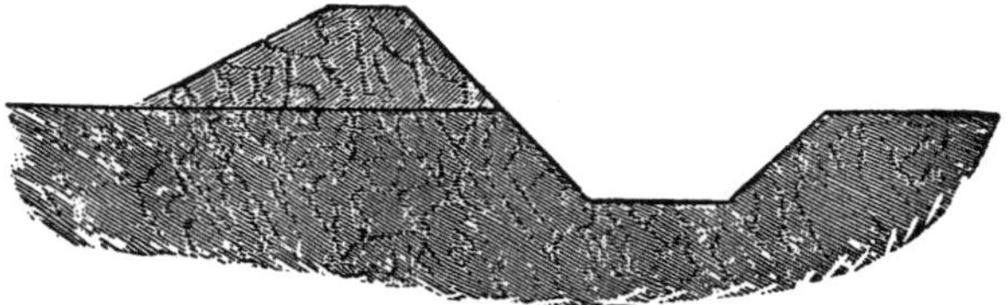

Fig. 17.—FORM OF FURROW.

a neighboring channel, while the lower bank is not raised, and permits the escape of a thin sheet of water over the ground below it. There are many forms of furrow available which will occur to the practical operator as they may be needed. But there are some methods of strengthening the furrows against wearing away by the currents

Fig. 18.—PROTECTED FURROW.

of water worthy of notice. One of these (shown at fig. 18), consists of a trough of wood, two strips of four or six inches in width being used. These are nailed together by their edges, and imbedded in the furrow. The water, in passing along, is prevented from escaping into or from flowing over the soil except at the open side of the trough. A portable wooden trough (fig. 19), with cross channels, may be used to convey water over ground that is under cultivation, or is not in a condition to be disturbed with the hoe. This trough is peculiarly adapted for use in the system of bedding before described, as it may be laid upon the crest of the head ridge, and the cross channels connected with the furrows upon the crests of the beds. These latter may be made of common open horse-shoe drain tiles inverted. The uses

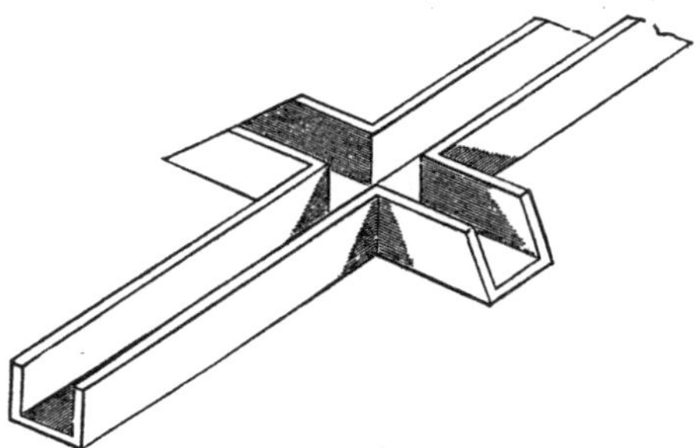

Fig. 19.—TROUGH FOR CROSS-FURROWS.

to which this kind of drain tiles may be put in surface irrigation are very numerous, but they will be so obvious to those interested that it is necessary only to suggest their usefulness in this regard.

For carrying the water beneath roads or paths a wooden pipe should be provided, (fig. 20). This is made of stout plank, placed longitudinally for the sides and cross-wise

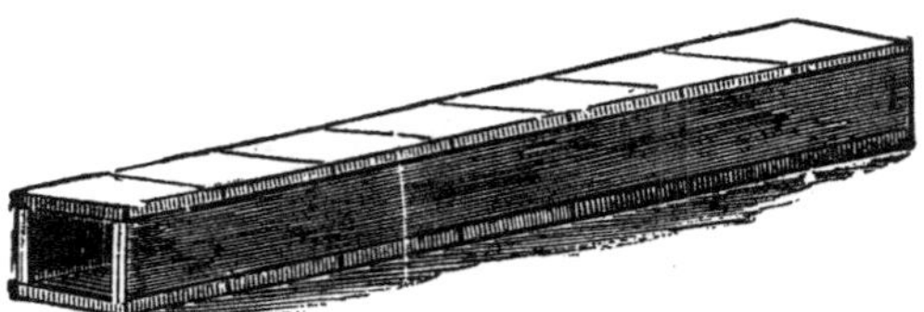

Fig. 20.—CULVERT FOR ROADS.

for the top and bottom. This method of construction gives the extreme strength of the material where it is most wanted, and prevents the crushing of the culvert by the weight of a loaded cart or wagon, the wheels of which might otherwise split the covering. These pipes are placed in the channels beneath the roads or paths, and the earth is heaped over them gradually, sloping in each direction. These pipes should be used wherever there is danger that earth may fall into the channels, or that they may be injured by rough usage. If made of seasoned oak plank two inches thick and bedded in waste lime or mortar, they will last many years without deterioration.

CHAPTER VI.

IRRIGATION BY PIPES AND TILES.

Many elaborate improvements have been made within the past few years in the practice of irrigation. The costly character of these improvements renders them inapplicable to any lands except those devoted to crops of great value. The minimum value of the crops that may be profitably raised by the methods of irrigation here referred to may be placed at $400 per acre. In some cases where the profitable use of land depends entirely upon these costly plans, this minimum may be reduced considerably. Thus, rather than have land idle it may pay to expend a permanent capital of $250 per acre, the yearly interest of which, with the annual cost of water, and labor, may on the whole result in a yearly outlay of $100 per acre, to produce crops which may realize $250 to $300 per acre. For a market garden these amounts are much less than the average value of the crops produced, and many seasons occur in which the losses by reason of dry weather at critical periods will amount to more than—or many times—the total value of the improvements here to be described. It is therefore a question of serious import to market gardeners, small fruit raisers, and the proprietors of private vegetable gardens, whether or not they could profitably adopt some of these methods which have actually been put in operation upon grass farms in England with very satisfactory results as to profit. From a careful consideration of this question, there will doubtless result a very decided opinion as to its feasibility and its profitableness. The simple fact that in many cases the crops which, under favorable circumstances, should have realized $600 to $1,200 per acre, have been so injured by drouth as to fail to pay the cost of production is sufficient to prove the propriety of this opinion, and to induce gardeners and fruit growers to adopt methods of securing a full crop in spite of the adversity of the season.

There are many cases in which the methods of surface irrigation previously described are unsuitable. Where the surfaces are irregular, where the crops are changed several times in a season, where the ground is under biennial or perennial crops and furrows cannot be maintained, or where the ground is too valuable to be occupied by furrows or water channels, these and other conditions will be favorable to the use of one or another of the following plans. The first to be treated of is that of underground pipes and stationary hydrants, from which water may be distributed under pressure through india-rubber hose and sprinklers. An elevated reservoir is provided, from which an iron pipe having a capacity equal to an inch and a half in area for each acre to be irrigated is carried along the center of the garden. A two-inch pipe will be required for two acres, a three-inch one for four acres, and a four-inch one for eight acres. From this other pipes are carried at right angles 200 feet apart to within 100 feet of the boundary upon each side. The pipes are laid a foot beneath the surface, or so far that they can never be disturbed by the plow, (see fig. 21.) Upon the lateral pipes, which should be at least an inch and a half in diameter, so that the flow shall not be unduly interrupted by friction, upright pipes or hydrants are attached which project at least three inches above the surface of the soil. These are about 200 feet apart. They are furnished with valves which operate by means of a square head and a key. Each one is fitted with a cap

Fig. 21.—IRRIGATING BY PIPES AND HOSE.

which screws on or off, and which is attached to the hydrant by a short chain for its preservation. When this cap is unscrewed a section joint affixed to the end of the hose may be screwed in its place.

When this apparatus is in operation, the water descending from the elevated tank or reservoir passes through the pipes and the hose, and escapes with some degree of force, depending upon the hight of the head, through a flattened nozzle, which scatters it in a thin sheet or broken shower. With this apparatus one man may water copiously five acres of ground in a day or night. Each hydrant being the center of a plot 200 feet square, serves to irrigate, with 100 feet of hose, very nearly or perhaps one acre of ground. To irrigate five acres in 10 hours would give an hour and a half to each plot, an amply sufficient time for an active man to get around a plot of 200 by 200 feet. The plan here described is illustrated in figure 22. The well, with reservoir, windmill and force-pump, is situated in the center of the plot to be irrigated at *A*. From this the pipes, shown by the double lines, are carried as has been described. The points

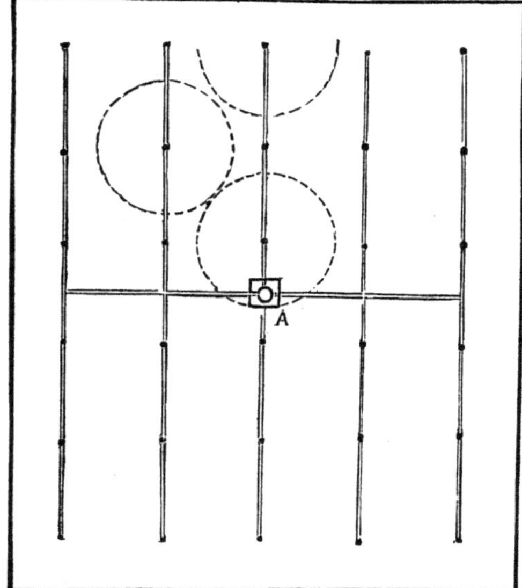

Fig. 22.—PLAN OF PIPES AND HYDRANTS.

marked upon the lateral pipes show the positions of the hydrants, and the dotted circles around a few of them show the extent to which the hose covers the ground.

A modification of this plan has been successfully introduced in England, where it has been patented, for the irrigation of meadows. In this method the distributing pipes are laid upon the surface of the ground, 30 feet apart, and are perforated in such a manner that the water is discharged in a shower of spray upon the ground, (see fig. 23.) This distance, however, will depend altogether upon the force with which the water is discharged or upon the amount of head given to the supply reservoir. The operation of this system is illustrated herewith. It has the disadvantage of increased cost, but the merit of economy of application. One acre is irrigated at a time and during one hour. The irrigation is done, as it always should be, at night, or between the afternoon and the morning. The apparatus is self-operating and needs only the turning on and off of the water by an attendant, who can be occupied with other work in the intervals. A plan

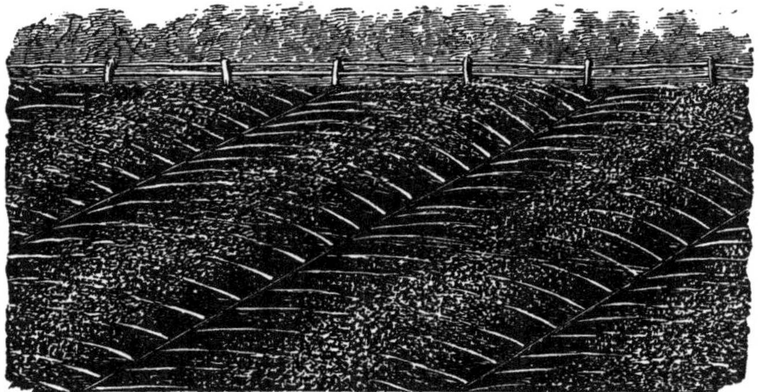

Fig. 23.—IRRIGATING BY SURFACE PIPES.

of subsoil irrigation by means of drain tiles has been in operation for many years, although a recent patent has been granted in the United States for the invention. The patent only refers to perforated tiles. But the common drain tiles will answer every purpose that the perforated pipes can or will. The plan is very simple. It is exactly the reverse of draining by tiles. Large pipes—the size being chosen to suit the system tributary to them—are laid down, a foot beneath the surface, at the highest part of the tract to be irrigated. From these, smaller pipes branch as the secondary channels of supply, and from them one-inch pipes again branch as distributing channels to the limits of the tract. The water escapes through the joints of the pipes, and rises by capillary attraction or absorption to the surface of the soil. As the water will naturally tend to sink in the soil in a greater measure than it will rise to the surface, the distributing pipes will need to be placed very closely, a distance of from six to eight feet being the greatest that should be allowed. This system has the advantages of cheapness of material, of permanence and of economy in applying the water. But it possesses the disadvantages of large cost of labor in laying the tiles, and of a very wasteful expenditure of water, a large portion of it escaping downward and useless to the crop. The trenches in which the tiles are laid may be very cheaply made by plowing twice or thrice in the same furrow until it is twelve inches deep, and when the tiles are laid, most of the earth may be plowed back into the furrow again. But one other objection will occur, in that for any sort of favorable result the slope of the ground must be regular, or the arrangement of the tiles must be made with costly exactness. Of the three systems here described, this last is the least promising, and should only be adopted in those special cases when, under a combination of favoring circumstances, it offers special inducements. Under such circumstances it has been successfully applied in California, and a correspondent of the "Rural Press," of San Francisco, from Santa Rosa, wrote recently to that Journal as follows: "I have practiced it on a small scale for several years. I lay down two-inch tile ten feet apart, so the top of the tile is just below the plowshare. I give thom just fall enough to run the water along, and fasten up the lower end. I make the entrance large enough to have plenty of head, than turn in a good stream of water that will force its way through. By this process the land never bakes, but keeps moist and loose. I believe one-fourth the water used under the ground is better than the whole on the surface." This opinion as to the economy of the practice will very probably be found premature on further experience.

CHAPTER VII.

IRRIGATION WITH LIQUID MANURE.

The ordinary cultivation of gardens exhibits a most striking want of economy. Farm gardens, and those smaller ones attached to village dwellings, ought to be cultivated in the most careful and economical manner. Not a drop of rain water ought to be allowed to go to waste. The house-slops should be carefully utilized. The cesspool, the stable, and the garbage-heap ought to serve the useful and appropriate purpose of aiding in the production of the household vegetables and fruits. But, on the contrary, it is doubtful if they are so utilized completely in any single homestead upon this continent. In some few cases they have been made to partially serve their proper purpose with the best effects. It is, however, in more densely-populated countries that liquid manuring has been practiced, and these valuable materials made serviceable. Without going so far as China and Japan, for examples of this economy, it may be stated that Belgium, the most thickly-peopled country of Europe, offers the nearest and most conspicuous example of the preservation of every kind of animal manure, both solid and liquid, and its manipulation in tanks for the purpose of applying its solution or dilution to gardens and small farms. In many parts of England, too, this system is closely followed, and the market farmers adjoining towns and cities carefully collect the waste of the dwellings for use upon their crops.

But in these instances, by reason of the abundance and cheapness of labor and the high value of the crops raised, rude and cumbersome methods of gathering, preparing, and applying these fertilizers are in use. It is very rarely that one can see even in England, in a small way, the thoroughly economic system of using liquid manures that are made use of in a large way for irrigating farms with the liquid waste or "sewage" of towns and cities. Their usual cumbersome methods are not adapted to our uses, yet we may gather from them some ideas applicable to our circumstances. There is, however, an arrangement of house drainage combined with garden irrigation recently introduced that has been tested with satisfactory results, and that is full of promise for its future general adoption. This grew out of the successful application of the system of earth closets to some cottages in a village in the county of Essex. The vast superiority of these over the common filthy cesspool, made more conspicuous than ever the inconvenience, insalubrity and waste of the usual slop holes where the liquid waste of the house was disposed of. For sanitary purposes a method was devised to dispose of this waste, and for economic purposes a plan of utilizing it was adopted.

From the sink of the kitchen a pipe, furnished with an air-trap, is made to discharge into a tank built of cement concrete outside of the wall of the house. The water from the roof is carried to the tank by a pipe, which also serves for ventilation. The tank is simply an above-ground cistern made water-tight and lined with hydraulic cement. The overflow from the tank is made intermittent by the ingenious use of a siphon, or bent pipe. The operation of this overflow is simple. When the cistern is filled to the movable cover, the water then trickles over the bend of the siphon into the drain. When this occurs the discharge of a pailful of water into the sink and through the pipe into the tank suddenly fills the pipe, flushes the siphon and sets it in operation, and the tank is drained to the level of the shorter leg of the siphon, The contents of the cistern flow away by a pipe, which leads from the drain. This tank is called the "self-acting flush tank." The cover is a movable plank floor which serves to allow access to the tank for any purpose. But this leads to the real subject matter in hand, the irrigation of the garden with liquid manure.

By this plan this can be secured whenever it is desired by simply introducing into the tank sufficient water to set the siphon in operation. The liquid than passes into the drain, and from that into subdrains of one-inch drain tiles placed one foot beneath the surface, and escapes through the joints of these into the soil. This arrangement is seen in the plan given in the accompanying illustration, fig. 24. The outer lines represent the boundary of the garden plot, supposed to be an eigth of an acre, or 50 × 100 feet. The tank is seen at *T*; the dark lines are the irrigating drains; the square dots are inspection wells, covered with a square stone or plank cover, by which examinations are occasionally made as to the condition of the drains, and the parallel lines between the drains are pipes which carry off any excess of moisture. This plan is capable of very extended application where the land to be irrigated may be beneath the level of the site of the house and the tank, and no house should be built on a lower level than the ground around it.

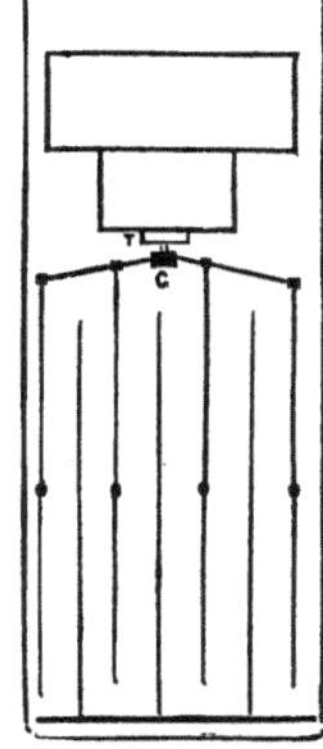

Fig. 24.—PLAN OF DRAIN.

An improved tank suitable for dwellings of a somewhat superior character is shown in figure 25. The principle is exactly the same as that previously described, the material of the tank being different. It is cylindrical in form, and may be of galvanized iron, of zinc, lead, or wrought iron, or of hard brick laid in cement. The discharge pipe may be of cast iron. This form of tank has been found to work with the greatest ease ; two quarts of water suddenly discharged into it when full being sufficient to set the flush into operation. This apparatus consists of the cylindrical tank, *A*, with a trapped inlet,

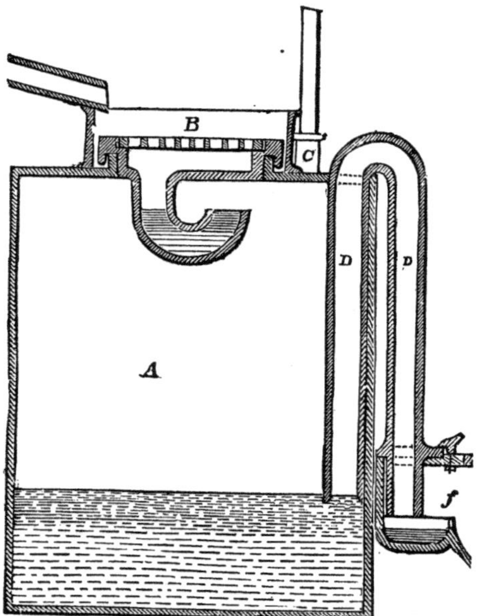

Fig. 25.—SELF-DISCHARGING SLOP-TANK.

which also forms a movable cover to give access to the inside of the tank. The pipe from the sink discharges over the grating of the inlet, *B*, as shown in the figure. A socket, *c*, is prepared for a ventilating pipe. There is also the siphon, *D*, and what is called the "discharging trough," *f*, consisting of a small chamber made to turn round, so that its mouth may be set in the direction that is required for connecting it with the line of outlet pipes, and provided with a movable cover for access to the mouth of the siphon. This "discharging trough" is an important feature in the tank, as it is of a peculiar shape, which, by checking the outflow of the liquid from the mouth of the siphon, enables a smaller quantity of liquid flowing into the tank to fill the bend of the siphon and set it fully in action.

In regard to the operation of this tank, and the drains connected with it, Mr. Geo. E. Waring, of Newport, R. I., writes as follows in the *American Agriculturist* of January, 1876: "I have found that in less than two minutes, about two-thirds of a barrel of liquid, (already accumulated in the tank), flows through the drain in a cleansing stream, which an examination shows to have left no refuse matters in its course. This tank is not yet made in America, and owing to its size and the cost of importing it, it is not likely that it will for the present come largely into use. In the meantime, the inventor has taken no patent in this country, and the invention is open to the use of all who choose to adopt it.

"The accompanying illustrations show how a perfectly efficient flush-tank may be made from a kerosene or other tight barrel without much expense. The barrel must be a sound one, with its bung well secured, and both of its heads in good order. Cut a circular hole in the upper head, twelve inches in diameter. Half way between the side of this hole and the chime, make another hole two inches in diameter. Finish the larger hole with an edging made of lead or copper, lapping over about an inch, and being securely nailed fast in a bed of white lead. This metal should be beaten in a groove or gutter just inside of the large opening, having its edge turned up at a distance of one inch from the edge of the hole. The head will then have an opening ten inches in diameter, surrounded by a channel three-fourths of an inch, or an inch deep, and one inch wide. A funnel of the same metal, (tin or galvanized iron would soon rust out), should be made to fit in this groove, its upper edge being turned over about an inch for the purpose. The funnel at its lower end should be furnished with a pipe turning up in such a way as to constitute a trap. Near the top of the

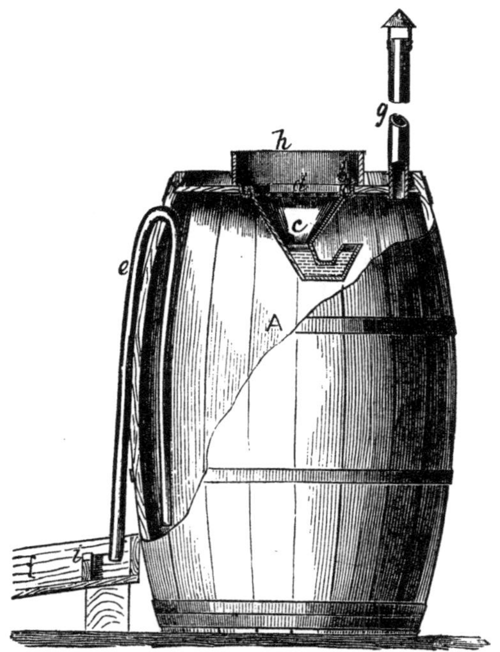

Fig. 26.—BARREL TANK.

funnel there should be a shoulder capable of supporting an ordinary round stove-grate of cast-iron ; this grate is intended to keep back any coarse matters which might obstruct the siphon, and to serve as a weight to keep the funnel in its place. Into the two-inch hole in the barrel top insert a ventilating pipe, which may be of tin, and which should be carried to the highest convenient point well away from any window or chimney top. Through one side of the barrel, close to the top, make a hole large enough to receive a $1^1|_2$ inch lead pipe, which, being turned down to within 6 inches of the bottom inside, and 2 or 3 inches lower at the outside, is to constitute the siphon for emptying the barrel— this pipe should not be larger than $1^1|_2$ inches interior diameter, as the larger the pipe the greater the amount of water needed to start it into action. The outer end of this pipe delivering into the drain is partially shielded from the access of air by an arrangement which will be described further on.

"Fig. 26 shows the arrangement of the whole apparatus. *A* is the barrel, *b* is the metal rim, or gutter surrounding the opening; *c* is the funnel with its trapped outlet; *d* is the iron grate; *e* is the siphon; *f* is the outlet drain; *g* is the ventilator; and *h* is a simple cylinder of galvanized iron or tin, to be used when the top of the barrel is above ground, so that it may be well packed around with leaves or litter without danger of these getting in to choke the grate. Where such packing is necessary, the whole affair should be housed in to protect it from the wind, and indeed it is always necessary to prevent the blowing in of rubbish which might plaster itself over the grate and prevent the water from entering.

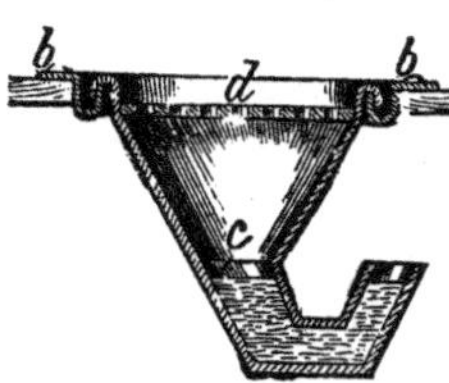

Fig. 27.—FUNNEL.

"Fig. 27 shows more in detail the construction of the rim, the funnel, and the grate. The gutter of the rim will be always kept full of water from the small amount splashing over, and this serves to seal the channel at this point just as the bent pipe at the bottom of the funnel seals its outlet. These seals are not liable to be forced, because of the ample air channel furnished by the ventilator.

"Fig. 28 is a longitudinal section, and fig. 29 a cross section of the outlet drain, show the arrangement for checking the flow of the siphon. A dam (*i*) which may be of wood, brick, or any other suitable material, closes the drain in front of the siphon to a hight a little above its lower end. This is notched down at its top to a point just below that of the end of the siphon, in such a way that after the barrel is discharged, the siphon itself will be emptied and will fill itself with air. This notch is too small to accommodate any considerable flow of the pipe, and the dam checks back the first water running, and helps to bring the siphon into action, but after the flow has all passed over, it lets the water behind the dam fall low enough to admit air to the pipe. I do not know that any thing further is necessary in the way of practical directions, except to say that the siphon pipe had better be attached to the side of the barrel, outside and in, by bits of tin tacked over it so as to prevent it from being injured. Indeed, the whole siphon might be inside of the barrel, its lower end passing out through a hole near the bottom ; this arrangement entirely obviates the danger of its becoming jammed, or the possibility of a trickling flow through it being frozen until the accumulated ice would quite close it."

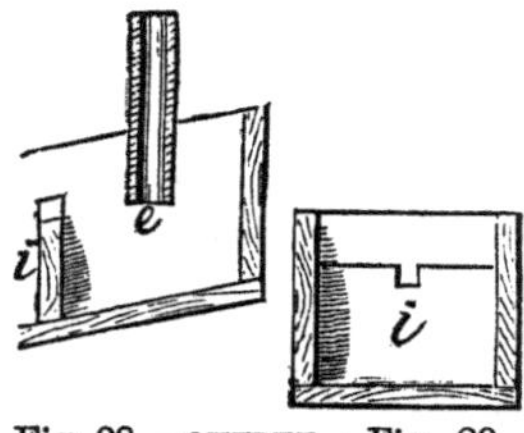

Fig. 28.—OUTLET.—Fig. 29.

The danger of filling up the pipes with sediment would prevent the application of this system to the use of matter from cesspools or barnyard manure tanks. It would not, however, prevent its use for the purpose of discharging a cesspool through a pipe of sufficient diameter, 4 to 6 inches for instance, into a manure tank in the stable yard, where it could be mingled with the liquid draining from the stables. This manure tank would then form the cesspool; the overflow from the house tank passing into it would flush and cleanse the latter at every considerable shower. A good supply of liquid matter of the very richest fertilizing power would then be at hand for use by means of permanent or temporary irrigating. The liquid would need to be raised from the cistern by a pump worked by wind or horse-power, as has been already described, and conveyed through large pipes into the distributing channels. These could be permanently made of inverted horse-shoe tiles, or in any of the methods heretofore mentioned, or temporarily by the use of the hoe.

In applying liquid manure it is always necessary to use it in a highly diluted state; even so much diluted that it would run off perfectly clear might be of sufficient strength for all purposes. The danger lies in using it of too great a strength rather than in diluting it too copiously. It has been found in practice when a heavy rain had filled the tanks at a season when there was but a very small supply of manure, and the dilution was certainly not less than a hundred times weaker than ordinary liquid manure, that the use of this weak liquid upon a plot of corn fodder, gave a wonderful stimulus to the crop, and the sudden change to an intensely dark green color proved that it was sufficiently strong, although from its color and freedom from smell the source of the liquid would not have been suspected. But it should be borne in mind that it is easy to injure a crop by using a too concentrated liquid manure.

For the most economical preparation and use of liquid manure proper cisterns need to be provided. The most convenient situation for these is the barnyard, where the drainage from the stables may be gathered, and where, above the cistern or near it upon one side, the manure may be heaped. When it is decided to use liquid manure there need not be so much attention given to the preservation of the solid manure, and although it may seem to be a sacrifice of this indispensable addition to the soil, yet it is far from being such in reality. On the contrary, the use of liquid manure is really an economy, and results in a saving of time and labor and increases the effectiveness of the solid manure. Being applied at the time when, and in the condition in which it will enter at once into the circulation of the plant, there is no loss of fertilizing matter. The crop, fed in its early stages of growth, receives its nutriment in such quantities and at such periods as will exactly meet its needs and force it into most luxuriant growth. In a dry season a plant may starve in the most abundantly manured soil; but when the manure is offered to it in a liquid form, and in copious supply, the growth is continuous and vigorous. A rapidly growing plant has the power to extract from the soil far more nutriment than a weakly plant possesses, and the stimulus afforded to a crop in its early stages enables the strong roots to penetrate far and wide in search of food, and the vigorous foliage is able to assimilate the abundant nutriment with rapidity as fast as it is supplied.

Every cultivator of the soil knows that a good start is the making of a crop, and this is precisely what is secured by liquid manuring. Therefore the solid manure may be used simply as the material from which to manufacture, by the help of all the needed rain or other water, as abundant a supply of liquid as may be. To extract all the soluble portion of the manure is the object, and what is left may be reserved to answer the purpose of top-dressing or mulching the soil in winter time, or of adding to its stock of slowly decomposing organic matter for future crops. It will be profitable therefore to adopt such a complete system of drains, tanks, and pumps, as will save every portion of waste from stable or manure-heaps, all the water that may fall upon the roofs and sheds, and occasionally pump up the contents of the cisterns and force them to filter back again through the heaps of decomposing manure. At the same time such accessory fertilizers as gypsum, the various ammoniacal salts, or soluble phosphates, or such deodorizing or fixing elements as sulphuric acid, largely diluted, may be added to the solution to increase its efficiency.

The construction of the tanks will be here the chief consideration. These, where the means are not ample, may be of the rudest character consistent with the ability to hold and retain water, but otherwise they should be constructed with a view to permanence and economy of use. A cheap and simple tank, of which a section is shown in fig. 30, may be made as follows: A pit or vat,

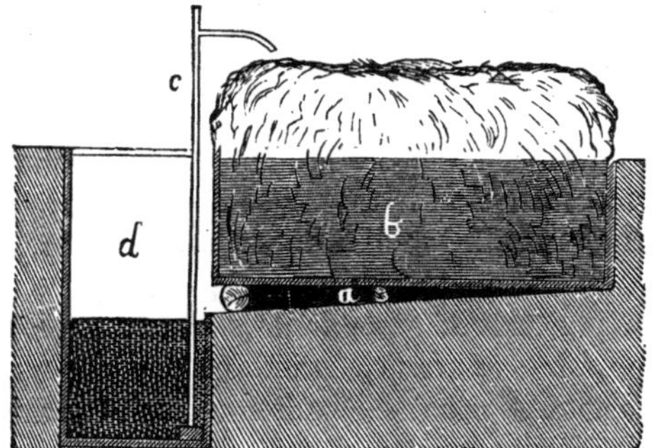

Fig. 30.—LIQUID MANURE TANK.

d, is dug and cemented with water-lime or lined with plank so as to be perfectly water-tight. This vat is covered with a plank floor, through which a wooden pump passes, and rests upon the bottom of the tank. The size of the vat of course will correspond with what is required of it. A useful size for a market garden, or for a farm where a few acres of soiling crops are raised each year, will be 16 feet square and 8 feet deep. At the end of the vat another excavation is made sufficiently large to contain the pile of manure or materials for a compost that can be gathered and used. This excavation, seen at *b*, may be

24 to 30 feet long, as wide as the vat, and gradually increasing in depth from 3 or 4 feet at the further end, to 6 or 8 inches more at the end connecting with the vat. The excavation should be floored with double boards, with a coating of asphalt or tar between them, and the sides cemented. A coarse grating of stout poles or timbers are laid across this shallow portion of the vat, and is supported in the center by blocks or short posts placed at intervals beneath it. Smaller poles or rails are laid upon these timbers not more than 6 or 8 inches apart.

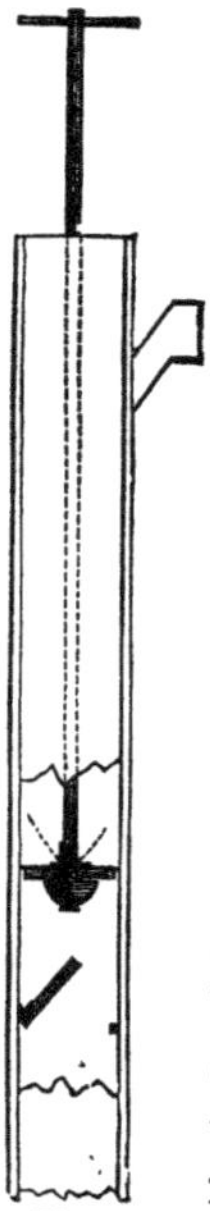

Fig. 31.
PUMP.

Upon these poles the manure is piled in a flat heap, made hollow or dishing at the top, so as to collect all the water that may fall upon it. The heap need not be more than five feet high, which is sufficient to cause an active fermentation to be kept up through the whole of it. The materials of which this heap is composed will include every thing of a mineral or organic character useful for manure, that can be procured. Stable manure, straw, marsh hay, weeds, sawdust, peat muck, leaves, woods-earth, night soil, leather scraps, tanner's waste, butcher's offal, lime, ashes, plaster, and bone dust, and the skillful operator will add from time to time such chemical substances as he needs to enrich the compost. There need be no fear of losing ammonia by adding lime. The lime is needed for the rapid decomposition of the manure, and the water added every day or two to the heap will seize upon every particle of ammonia formed and carry it into the tank, where it may be fixed by the addition of sulphuric acid or gypsum. The water in the vat should be frequently pumped out for use, and a fresh supply poured upon the heap. A pump that will not readily be choked should be used.

One with a collapsing bucket, with leather sides and of a conical form, shown in fig. 31, is the most useful. The waste water from the roofs might be discharged upon the heap by a simple arrangement of spouts. The object desired, viz., to gather every soluble part of the manure into the vat, should be forwarded by every possible means.

A small and cheap tank suitable for use where the liquid manure from the stables and dwellings may be collected for distribution, may be excavated and lined with brick, and should be circular with an arched roof; if made 12 feet deep and 10 feet in diameter, it will contain 6,500 gallons, or sufficient to irrigate an acre with nearly three-fourths of a quart to every square foot.

Many other forms of tanks may be used for this purpose. A capacious one may be constructed as follows: A circular pit 24 feet in diameter and 8 or 10 feet deep is excavated. The bottom and sides may be cemented or lined with bricks laid in cement. A pillar of brick is built in the center, and a brick arch may be sprung from the pillar to the wall around it, or beams may be laid from the wall to the pillar, centering at the pillar, and a plank floor may be laid above them. A wide spout or throat leading from the manure heap may carry the liquid into the tank, and the drain pipes from the stables and dwelling may be made to discharge into it. An engraving of this tank is given on page 37. Many obvious modifications of this plan will occur to the reader.

The distribution of liquid manure may be made, as already described, through pipes or open furrows, or by means of irrigating carts or barrows. The use of carts will be found to require a very small outlay at the beginning, and to be much more satisfactory than would appear at first sight. Where the distance to which the manure has to be carted is within 400 feet, which would be from the center to the outside limits of a square of about 15 acres, carts would be the cheapest method of applying the liquid. One acre an hour could be easily watered by a one-horse cart, furnished with a spreader six feet long, that would cover a width on the ground of about eight feet. If the crop is grown in drills two feet apart, the horse would occupy one drill, each wheel of the cart one drill upon each side, and the spreader would cover half a drill upon

Fig. 32.—LIQUID MANURE CART.

each side; thus four drills would be watered at each passage. If the drills are three or four feet apart, three or two are watered at each passage. In this way one acre would be watered at an expense of about 50 cents, allowing $5 for the cost of the necessary horse, two watering carts, and two men. Two carts are needed, as one would be filling while the other is spreading. The cart may be a large cask holding at least 200 gallons, mounted upon wheels with a pair of shafts, the axle being bent to keep the load low down, and a distributing pipe perforated with holes, and curved forward at the ends. Instead of a barrel a square tank mounted upon wheels may be used, see fig. 32. The supply is regulated by means of a ball-valve attached to a wire, which is pulled by the driver when the valve is to be opened, this is shown at fig. 33.

For the irrigation of smaller gardens a hand-barrow, with a distributor as shown at fig. 34, or with a small force-pump and sprinkler attached, would be useful. It would serve in cases where no more than an acre can be appropriated for garden and for fodder crops to support a single cow or horse, as in thousands of instances which occur in village dwellings or the suburbs of cities and towns. By cultivating small tracts of an acre or less upon this system, the domestic supply of vegetables

Fig. 33.—VALVE FOR DISTRIBUTING TANK.

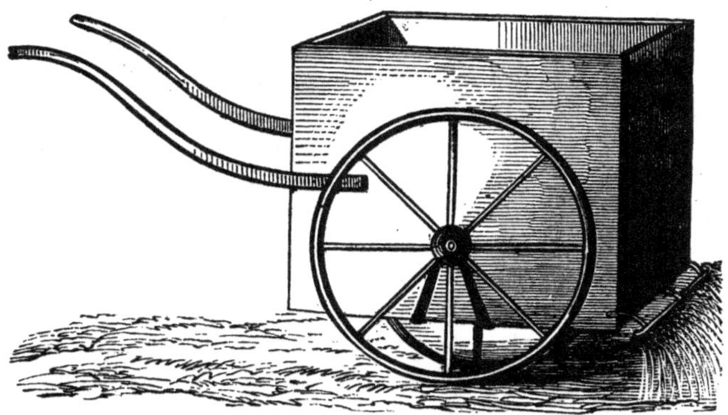
Fig. 34.—HAND-BARROW.

may be easily raised, together with ample support for one cow. It is the vast number of such cases as this to which a system of irrigation may be applied, that an aggregate of benefit may be derived that will almost balance in individual comfort and advantage the more conspicuous but less numerous systems of field and farm irrigation. There are probably two or three millions of individuals to whose small occupations this system may be applied with a result equaling at least a net amount of $100 per year, in each case. The material that might be utilized, in most cases now goes to waste or serves to vitiate the air and poison the surrounding neighborhood. By thus turning it to account, a benefit to the public health incalculable in dollars and cents would result, and at the above reasonable estimate a vast addition to the wealth and comfort of the people besides.

A plan for utilizing liquid manure upon gardens or farms so situated as to surface that the manure may be spread by gravitation, or flowed in furrows from a drain issuing from the barn cellar, or from a tank in the barnyard, was described by Col. G. E. Waring, jr., of Ogden Farm, Newport, R. I., in the *American Agriculturist,* September, 1873. Mr. Waring taking his cue from the method of utilizing the sewage of towns upon some English farms says: "A very important lesson for many American farmers may be gleaned from the English experiments in the use of sewage as manure.

"Mr. Mechi, a well known English farmer, still adheres to his old system of converting his manure (or much of it) into a liquid form, storing it in a large tank where it ferments, and forcing it (by steam-power) through underground iron pipes for distribution over the land through a hose. This system is not generally considered either economical or advantageous. The plan adopted with sewage, in all cases which came to my notice, is that described as in use at Lord Warwick's farm near Leamington.

"While our climate precludes the possibility of using winter sewage in this way, we might, in some cases, make profitable use of summer sewage if we could get it without too much cost. What most interests us in the matter, is the suggestion that we may adopt a similar means for the use of water as a distributing medium for manure.

"I will take as an example my own case at Ogden Farm, and will assume that I had (which is not true) a stream of water at a sufficiently high level to be led into the barn cellar (40 × 100), which has a capacity of about 200,000 gallons. This should ordinarily be kept nearly full of water, and into it all manure should daily be thrown. Care must be taken to ventilate the cellar thoroughly with side windows, and to have the stable floor above it quite tight. An arrangement should be made to turn the stream into the cellar, or back again into its own channel at will. Whenever manure was required for that part of the farm lying low enough to be flooded from the cellar (about one half of the whole), the gate should be opened and the liquid conducted to the field by the system explained below. At the same time enough water should be admitted from the brook to keep up the head in the cellar. This, by its flow, would make a movement in the mass sufficient to stir up the sediment and foul the outgoing water. The irrigation should be as frequent and as copious as the supply of water would allow, and as the best growth of the crops required. The water alone would be very beneficial, and it would only be stronger or weaker according to the extent to which it was employed. Of one thing we might be quite sure; all the manure it contained would be distributed in the most perfect way possible, and there could be no waste. The water would be an *addition* to its value—there would be no deduction in any way. A vast amount of labor would

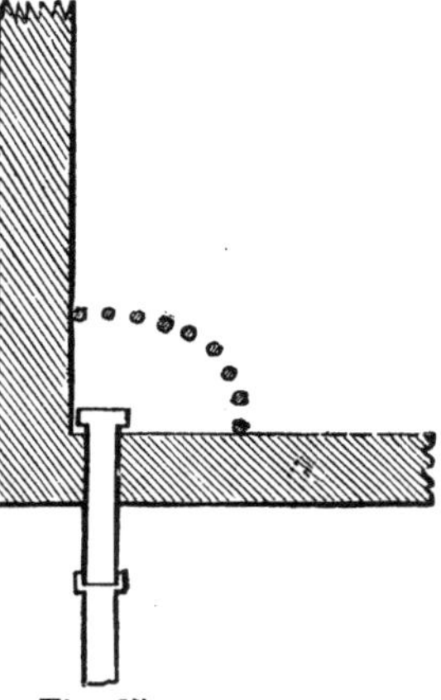

Fig. 35.—ESCAPE PIPE.

be saved, and the manure would be applied *at the right time, in the right way, and on the right spot.*

"The winter manure should be hauled, as it now is, on the higher parts of the farm—no water being admitted to the cellar at this season. When the growing season came on, then the crops of the lower parts would get the benefit of the irrigation. How great a benefit this would be to grass land in time of drouth need only be suggested.

"The accompanying sketches will show the arrangements to be made at Ogden Farm, and will indicate a

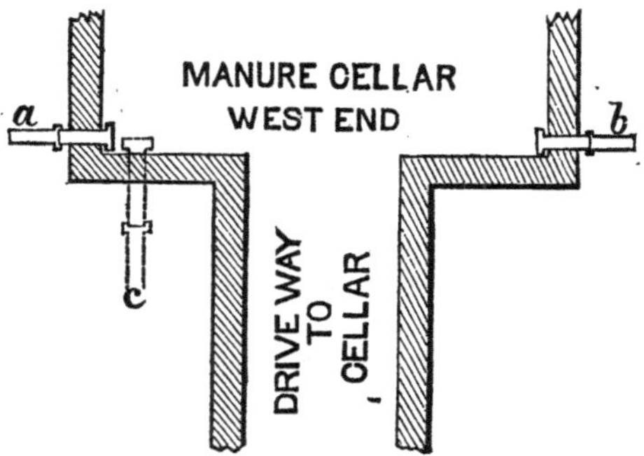

Fig. 36.—DRIVE-WAY.

plan which, with such modifications as circumstances require, may be adopted for the irrigation of any land with sufficient slope.

"Fig. 35 shows a corner of the manure-cellar with an escape pipe (valved) leading from the very bottom—allowing the cellar to be drained dry at pleasure. In front of the entrance to this pipe a screen of iron rods or wooden slats, reaching vertically from floor to ceiling, prevents solid matters and litter from choking the pipe. If this becomes clogged, it can be cleared with a rake through a trap-door in the floor above. This pipe should be used only when the water will not flow at the outlets above.

"Fig. 36 shows the arrangement at the west end of the

cellar, with an overflow pipe to the north and one to the south. The drive-way should be dammed up to raise the water to the level of these pipes.

"Fig. 37 shows the arrangement for the distribution of the flow. A main furrow runs from *a* and *x* to *d*. This

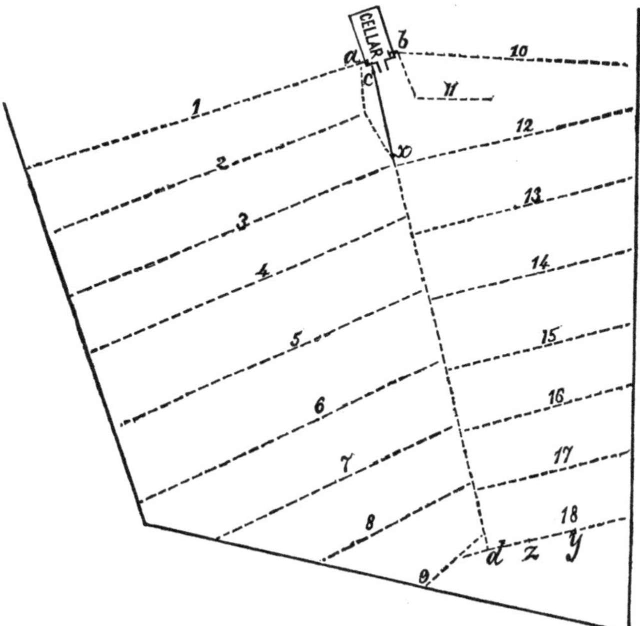

Fig. 37.—PLAN OF FURROWS FOR DISTRIBUTION.

is the general direction of the slope of the land. The laterals 1 to 18 are furrows laid on a fall of 1 inch in 100 feet. They will not be straight, but must follow the conformation of the ground, so as to preserve a uniform fall. The main furrow at *x* may be supplied either from *a* or from *c*, and others from *b*, as in figs. 36 and 37.

"The flow being let on, and kept up by a corresponding flow into the cellar from the brook, it should pass on to the end of 18. (The main furrow is a little deeper than the entrance to the laterals.) Here it will overflow the land lying below so much of the lateral as is beyond *y*. Then a gate should be set at *y*, and kept there until the land below the lateral between that point and *z* has been sufficiently flooded. Then remove the gate to *z*. When all the land below lateral 18 has had its supply, set a gate in the main just below 17, and repeat the process with that. When the south side of the farm has been completed, the gate is taken from the main and the water allowed to flow to the end of No. 9.

"Nos. 1, 2, 10, and 11 can be flushed only from outlets *a* and *b*. All the others are low enough for *c*.

"Of course, any portion of the land may be flooded at pleasure, the directions above being given only as an illustration."

The scope for the employment of such methods as these suggested in this chapter is far from narrow. The profitable employment of liquid manure upon gardens and small farms upon which the crops grown are of high comparative value, cannot be doubted. It remains only that the lead in introducing it be taken by some enterprising but cautious man, in each neighborhood, whose success would stimulate hundreds of others to follow his example. It is probably too soon to more than hint towards the use of liquid manure upon farms in this country, or the utilization of the sewage matter of towns and cities. This can only be done with profit when the high value of lands bears some proportion to the cost of the necessary machinery. But upon gardens, especially market gardens, and upon highly cultivated farms where heavy fodder crops are grown, and the soil is abundantly manured, and where the closest economy in the saving and use of manure is practiced, much may be done in this way. The author has had practical experience in the use of liquid manure—in gardens, and in growing fodder crops, to be used for soiling dairy cows—and is firmly convinced that, with ordinary care and ingenuity, the crops may be quadrupled, and the profit doubled. For instance a clover crop that would under ordinary circumstances be ready to cut for soiling only in June has, by weekly irrigating with liquid manure, been made ready early in May, and by more frequent watering has been cut four times before the first of July, or once every two weeks after the first cutting, at a cost, for each watering, of not more than 50 cents per acre. Each cutting of the crop at least equalled an ordinary yield, or one ton and a half of hay per acre.

As to the value of the system as applied to market gardens for the production of such crops as onions, cabbages, cauliflowers, and the smaller vegetables, in which flavor, tenderness, and succulence are only secured by rapid growth, there can be no better proof than the successful cultivation of the small farms of Belgium, a country which supports the densest population in Europe, or of the market gardens in the vicinity of many French, Italian and English cities and towns. In these localities the solid and liquid refuse is gathered with the greatest care, mixed so as to be readily used, and applied to the crops, which, under this treatment, possess a size and quality that is never equalled in this country, except by a few premium vegetables that are grown in this same manner. To have seen this demonstrated in the gardens and in the markets of European cities and in isolated cases in this country, is sufficient proof, at least, to induce American cultivators to *attempt* to utilize in this most effective manner this most effective fertilizer.

CHAPTER VIII.

CULTURE OF IRRIGATED GARDEN CROPS.

There are a few important leading principles involved in the practice of irrigating gardens that should be well considered. These will be referred to in the order of their importance.

Drainage.—It is rarely that a well drained soil can be injured by a copious supply of water; but one that is not drained may easily be turned into a quagmire by an excess of it. Drainage, therefore, should be the first thing provided before this method of cultivation, let it be complete or partial only, is attempted. If the soil is not naturally drained by means of an open and porous subsoil of sand or gravel, tile drains should be laid in such a manner as to carry off the surplus water in the most effective manner.

The method of drainage will depend upon the system of irrigation adopted. If the bedding plan is used, as illustrated in fig. 7, page 42, the drains should be laid between the beds, and beneath the drain furrows, as shown in fig. 37, in which the open spaces seen at *a, a,* represent the drain. These drains should be of inch tile, laid three feet below the surface. If laid at a less depth there is danger that the roots of some varieties of plants may penetrate between the crevices and choke the tiles. Where the arrangement of the water-furrows is such as to need change every year, or such as is shown in figures 15, 17, or 23, the method of drainage should be the ordinary one of inch tiles laid 24 feet apart, if the soil is heavy; or 30 feet if of a lighter character, and leading into main drains of two or three inch tile. Surface draining would be a very unsatisfactory resource, and should be adopted only where the crops would resist the effects of a very cool, moist soil, or upon inclined ground where there would be no danger of saturation. Whatever the arrangement of water supply may be, the plan of the drain should be as nearly as possible exactly the reverse. In effect the drains should be so arranged as to take up the surplus or unused water and carry it off as rapidly as possible; at the same time care should be taken not to permit the water to flow into the drains until it has done its duty, nor to use so much water that the soil may be carried into the drains and these be soon filled with sediment. No drain should be carried beneath a canal or distributing furrow, unless it cannot be avoided, and then never at a less depth than three feet, else a channel of communication may be opened between them and the water escape, and, what is worse, wash the soil into the drains and render them useless. Further remarks upon drainage will be found in a succeeding chapter where field irrigation is treated, and which may be referred to.

Cultivation or Disturbance of the Soil.—The soil should never be disturbed while it is wet. The operations of hoeing, cultivating, weeding, sowing, or gathering the mature crops, should be so timed with reference to the watering, or the watering should be so timed with reference to them, that these operations may be performed when the soil is dry and just before the watering. If after the watering, upon soils liable to "bake," or become encrusted, the surface under the effects of a hot sun becomes hard, the crust should be broken up by cultivation before it has time to completely harden.

The Application of Water.—It is not well to put off the watering until the ground is very dry, but to apply the water while the soil is still somewhat moist and mellow; it is then more absorptive, and the after effects upon the worst of soils, as regards baking, will be less troublesome. The soil should be moderately watered a day or two before seed is sown or plants are transplanted, that it may be in a finely pulverulent condition, and when the supply of water is always under the control of the operator there is no danger in sprouting the seed and thus hastening germination. After sowing or transplanting, the chief care should be to water only very moderately, and never allow the water to flow over the seed or plant rows, lest the surface should become hard and need stirring, and the young plants be endangered by one or the other of these alternatives. Moderate, frequent waterings are best for young, growing plants. There is far greater danger of giving too much rather than too little water at this time. During early growth the application of water at a lower temperature than that of the soil is injurious. For this reason, when well-water is used, it should be exposed to the air in open tanks or reservoirs for at least one day before it is used. For the same reason watering during a clear sunny, or a windy day is to be avoided, and it should only be done in the evenings, or when the sun is obscured with clouds. The effect of wind is to increase the evaporation, and thus reduce the temperature of the soil immediately after its saturation. The quantity of water to be applied will depend upon several circumstances that have already been referred to. For garden crops, frequent moderate waterings are preferable, and intervals of five days are usually allowed. The soil is then kept constantly moist, and the growth of the crops continuous. Of course when rain falls, a sufficient allowance must be made, but, judging from the quantities of water that may be safely applied to crops in the market garden, unless the rain is unusually heavy and continuous, it may safely be ignored. The quantities used in garden culture in different countries, as mentioned in many works upon irrigation, are exceedingly irregular. It would seem as though the abundance of water, and the porosity of the soil, measured the supply, rather than the needs of the crop. Thus quantities varying from a total depth, during the growing season, of 50 up to over 300 inches upon the surface, have been used without any ill effect when the drainage has been perfect. Experience can be the only safe guide; the thorough soaking of the soil at intervals of five days, should be the limit of the irrigation, and the quantity of water needed to effect this will be the maximum supply required. When economy of water is a point to be considered, as it must needs be when every pint of it is elevated by power, it will be necessary to watch the flow in the distributing furrows and prevent any escapes into pools and surface drains, and such copious watering as would leave water standing in the furrows for more than an hour or two after the flow has been stopped. This must be regulated by the judgment of the irrigator acting through a knowledge of the principles involved.

THE MANAGEMENT OF VARIOUS CROPS.

Where the climate admits of it a succession of crops of garden vegetables may be grown throughout the year, and the variations of the seasons practically removed. In the climate of California this is easily done by means of irrigation there practiced, and in most of our Southern States the season of growth may be extended, and in some may be continued throughout the year, if the supply of water is only secured. This is one of the great advantages of a system of irrigation, by which every where a succession of crops, more or less extended, may be secured. The general management of the principal garden crops will be briefly indicated.

Asparagus.—The most convenient method of cultivating this crop is by "floors," (see fig. 9, p. 43), over which a thin sheet of water may be flowed from a furrow at the head towards another at the foot, from which the water may be again flowed over another floor below the first.

This arrangement makes it necessary that the ground should slope slightly in one direction. The method of watering by pipes laid upon the surface, or by hydrants, which have been already described, may easily be applied to the culture of this vegetable. This crop is one that needs but very moderate irrigation.

Beans.—This crop requires to be planted in beds, arranged as shown in figs. 7 and 8, and can be cultivated in long succession by means of irrigation. It will stand a good deal of moisture, especially when grown to use green as "snap beans" which should be fresh and succulent. The periods of irrigation should be at intervals of five to seven days. Lima beans need equally frequent waterings.

Corn.—This is a plant which needs much moisture, and the watering may be both copious and frequent. It may be planted in hills or drills, in either case the system of beds or of alternate drills and furrows, which are fed from a distributing canal at the head of the bed or drills, may be used.

Cabbage.—This crop is cultivated in beds to which the water is supplied by furrows, made with the hoe after each cultivation. It is a greedy feeder and responds quickly to the application of liquid manure. Heads of enormous size have been thus grown, and specimens of 60 pounds in weight have been frequently exhibited that were produced by irrigation with liquid manure. It will submit without complaint to much moisture if the soil is cool; how it would behave under our hot suns, when stimulated by excessive irrigation, is something that is yet to be learned. In Florida, however, it thrives well when supplied with sufficient moisture; in central Europe, where the market gardener irrigates all his crops, the cabbage is only moderately watered, doubtless lest it might be stimulated to run to seed; but where the character of the soil and climate are favorable, and abundance of water is procurable, there the cabbage, as well as the cauliflower, is extensively cultivated not only for home consumption but for shipment abroad to distant countries. This is the case in Belgium, and in the neighborhood of Erfurt, (Germany,) where both of these crops are cultivated with success and profit, unequaled elsewhere. There the method of culture is to choose a low spot of ground and divide it into beds of convenient shape, which are separated by permanent furrows, in which the water flows. The water is sometimes dipped from these furrows by long-handled scoops and poured around the roots of the plants. Otherwise the water is flowed on to the crops by means of small furrows between slightly raised ridges upon which the plants are grown.

Beets.—This crop is peculiarly suited to culture by irrigation. Few crops thrive so well under the combined influence of abundant moisture and a continued high temperature. The sugar beet, especially, enjoys these conditions when planted in deep, well-drained soil, and crops equal to from 60 to 75 tons of roots per acre are frequently grown in the sugar manufacturing districts of central and southern France. A specially noteworthy case was cited in the *Journal d'Agriculture* by M. Barral, in which a manufacturer of beet sugar at Masny, directed the flow of water from the water wheels, which furnished the power for the factory, on to the field of beets. The water was charged with all the refuse of the works, the washings of the roots and of the impure bone-black, as well as that of the sacks in which the pulp had been pressed, the skimmings of evaporating pans, and also the washings of the outhouses used by the workmen ; and carried all this matter in suspension through the channels and distributing furrows to the growing crops. No other fertilizer has been used during 8 years, and the value to the farm is estimated at a yearly sum of $2,000. This example, however, relates to field culture, but is yet worthy of note as showing how refuse matter may be applied in a similar manner to garden crops. The irrigation of beets, although it may be profusely applied upon light, deep, and well drained soil, must be done with proper moderation upon soils that are retentive and not well drained. Only so much water must be used as to keep the soil fresh, moist, and mellow, and it will be safest to irrigate such soils as these more moderately and oftener than those of a loose, open, sandy character.

Carrots.—This crop has been found to thrive exceedingly well under irrigation upon light soils. A succession of crops may be grown throughout the whole summer, and by the use of some active artificial fertilizer, the growth is rapid and remarkably clean and healthy. Upon clay soils this and other deep-rooted crops do not thrive very well, and more shallow-rooted crops should be chosen. When irrigated, the carrot is cultivated in rows upon the flat, the water being led to the plants in channels made by the hoe in the intervals between the rows. It is very common in garden culture to plant carrots for a late crop in rows between other and earlier ones, by which the tender young plants are shaded and protected from the heat.

Sweet Potato.—This crop is planted in broad, flat beds slightly raised above the level, and the water is flowed into the furrow between the beds. Upon the light soils, in which the crop succeeds best, the waterings are given copiously at intervals of from five to seven days. The abundant foliage requires a good supply of water. The system of rounded or doubly sloping beds, described on page 41, in which the water is carried along the crown of the bed, is well adapted to the culture of this root.

Onions.—This crop is grown very successfully under irrigation, and water may be copiously applied. The excellent quality, mild flavor, and extraordinary size of the

Portugal and Italian onions are due to their manner of growth in which irrigation is extensively used. The crop should be planted in rows between which water is flowed, in broad, shallow channels made with a hoe. The water should not come in contact with the bulbs, nor should the earth be thrown upon them in making the furrow.

Potatoes.—To grow common potatoes under irrigation, with success, needs caution and judgment. As the quality of the tubers depends greatly upon the supply of water, judiciously regulated with regard to the character of the soil, some care must be exercised as to the quantity. Upon light soils the water is given only at intervals of nine or ten days, and upon heavier soils, which are more retentive, fourteen days elapse between the waterings. As soon as the soil is sufficiently dry after watering, the surface should be cultivated, which will cause the moisture to be better retained. A system of drills, or of beds slightly raised, is used for this crop, the water being given in broad, shallow furrows, made with the hoe at the time of cultivation. When the plants nearly cover the ground, as they should do at the time of blossoming, the final watering is given. No further cultivation should be given after this period.

Peas.—As this crop is generally sown in rows upon a flat surface, the mode of watering should be suited to this method of planting, and it may be either by a system of beds, fig. 6, or of shallow furrows made between the rows with the hoe at the time of cultivation. As this crop flowers and seeds during a lengthened period, it may be irrigated without regard to the flowering, care of course being taken to keep the soil only in a healthful state of moisture.

The Small Crops.—Small crops, such as lettuce, radishes, etc., are more conveniently cultivated in beds of the form shown in fig. 6, over the surface of which the water flows or trickles from a furrow at the ridge. The quality of all these small vegetables is improved by copious waterings, and a very profitable succession may be procured by continuous sowings, the growth of which for market or domestic use may be hastened and matured at pleasure.

GARDEN FRUITS.—The various small fruits usually grown in gardens may be greatly increased in luxuriance of growth, and by cautious treatment, much improved in quality, by irrigation. Over-watering, however, will infallibly tend to deteriorate the quality, if it does not even weaken the growth. As soon as the blossom appears water should be withheld, unless under extraordinary circumstances, and under the supervision of an experienced gardener. For strawberries the bedding system is preferable, and for other fruits the water may be led by temporary furrows made with the hoe around the roots of the bushes or the vines.

In concluding these remarks which are not intended as a guide to an already practiced and competent gardener, but as suggestions to those who desire to secure in a moderate way by the use of some plan of irrigation, that is feasible for them, the full advantages which they can derive from a family or market garden, and which they so often fail to gain, by reason of the frequently recurring drouths ; it may be said as a matter of caution, that with a supply of water constantly at hand, the danger of using too much is greater than that of using too little ; that moderately copious waterings at extended intervals is far preferable to light but frequent irrigation, which scarcely reaches the roots and packs the surface. To saturate the soil once a week, or every ten days, will have the effect of forcing out of it much of the air that is contained in it, which will be replaced by a fresh supply as the moisture evaporates or sinks in the subsoil. Thus the soil is kept loose and mellow, and the necessary cultivation, which should always follow the watering, will retain this condition of the soil. The crops then are refreshed and invigorated, and can resist a comparatively long interval of dry weather. An excess of water may very easily be worse than a severe drouth, for permanent and irreparable injury may be done to a crop by flooding the soil to excess; and not only the season's crop itself be lost, but the plants themselves be seriously damaged and future crops be imperiled. With caution in this respect, an adequate sonsideration for the peculiar character and needs of the different plants, a sufficient regard for the nature of the soil and its facilities for proper drainage, whether natural or artificial, and some reference to the ordinary provisions of nature in regard to the supply of water, one can scarcely go wrong in applying the practice of irrigation to the culture of any of our usual crops of vegetables, fruits, flowers or shrubs. The general application of irrigation, with few exceptions in this country, will be to make up for the short-comings of dry seasons, in which the deficient supply of rain may be made up artificially.

CHAPTER IX.

IRRIGATING ORCHARDS AND VINEYARDS.

It is doubtful if there is a single orchard or vineyard in the United States, except in California, Utah, or Colorado, subjected to a systematic irrigation. At the same time it is doubtful if there is any country in the world in which irrigation could be more profitably applied to fruit culture than here. The experience of orchardists proves that drouth is accompanied by destructive attacks of insects. How far these depredations might be prevented by irrigation cannot be predicated, but it is beyond doubt that the vigor of growth that would result from a sufficient supply of moisture to the roots would greatly mitigate the effects of these attacks. The apple trees that never have an "off-year" are those grown near bodies of water. A California vineyardist who irrigated his vines immediately raised his product to eight tons of grapes per acre, and greatly improved the quality. The newly planted orange groves of Florida are frequently destroyed by drouth, and methods of irrigation are eagerly sought to render their culture more safe and certain.

But if it were necessary to enforce the advantages of the irrigation of orchards, abundant evidence could be gathered in the south of France, Italy, and other countries of Southern Europe, where the olive, orange, lime, almond, fig, apple, and other orchard trees, as well as the vineyards, are systematically brought under irrigation. As to the vine, it is a question which so far has not been thoroughly investigated, whether or not irrigation might be made the means of vanquishing the destructive phylloxera. An experienced vineyardist of Avignon (France) submitted his vines during the Winter, which in that locality is mild and free from severe frosts, to a lengthened irrigation of 30 days, during which a depth of four inches of water was constantly maintained in the vineyard. This operation has been found to considerably diminish the injurious effects of the phylloxera, and to greatly improve the condition of the vines. This practice might be found somewhat dangerous where early Spring frosts occur, by which the vines brought prematurely into growth might suffer. But no cautious cultivator will make serious innovations upon his practice without previous careful experiment. In Southern California the vineyards are copiously irrigated four times only—at the starting of the first growth, at the blossoming, at the setting of the fruit, and at the period when the fruit commences to color.

But without entering into speculations as to what events might occur, it is sufficient to know that orchards are irrigated with profit; that in some cases they are destroyed, and in numberless instances they are injured by a want of water, and that there are probably few cases in which a supply of water brought into the orchard would not be found advantageous and profitable. The methods of irrigating orchards are very simple. It is only necessary to put the water where it will do the most good, and that is as near as possible to the extremities of the rootlets. The extent of the roots of a tree bears a ratio somewhat approaching that of the branches. Near

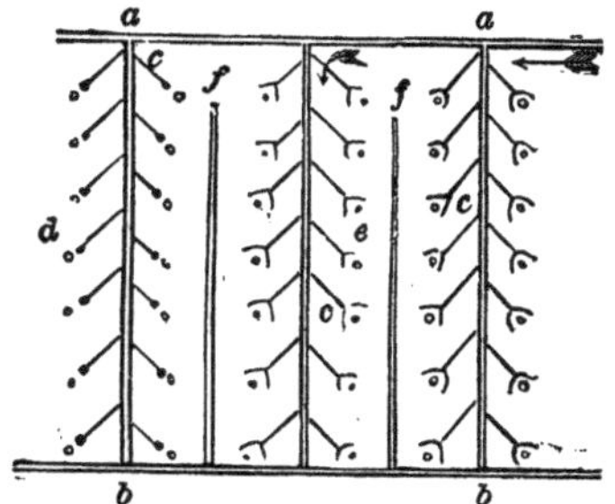

Fig. 38.—PLAN OF IRRIGATING AN ORCHARD.

the stem there are few of the root-hairs or fine fibers by which nutriment is absorbed. These are found at the extremities of the very fine rootlets, and these exist in a ring around the tree, the inner edge of which is from 3 to 4½ feet distant from the stem. In irrigating an orchard, then, the most perfect method of applying the water is to distribute it in a broad circular channel around the tree, distant about six feet from the stem.

Where irrigation of orchards is practiced two different plans are adopted. The first is a somewhat rude method, but is easy and effective. The water is led into a channel between two rows of trees, *a*, *b*, fig. 38, and from thence into distributing canals, *c, c, c,* which carry the water within a few feet of each tree. (The position of the trees in the figure is indicated by the dots.) Here a sharp bar is trust into the ground in several places, penetrating in different directions toward the roots, and leaving holes by which the water soaks into the earth and reaches the roots. The second is a more elaborate but a more preferable method. The water is led from the canals into circular furrows which curve so as to embrace the tree. (This is shown at *d, e,* in fig. 38.) These furrows are broad and shallow, and the water overflows from them in a thin sheet or a multitude of little rills which lead to

Fig. 39.—FORMATION OF FURROWS.

the lower side of the tree, where they are arrested by means of a slight embankment raised with the hoe. In this case the water is brought exactly where it is needed, and every rootlet is supplied. This is also seen at fig. 39.

In irrigating vines very similar methods are adopted. As the vines are planted in rows, the distributing furrows are carried down the center of each alternate row, fig. 40, the ground being sloped towards the center of each intermediate row, fig. 41. The water is thus made to pass across each row of vines. Beneath the center of the intermediate rows a tile drain should be placed to carry off surplus water, and this brings into notice the question of drainage as a part of this system of orchard and vineyard irrigation. As a rule irrigation and drainage should go together. Irrigation without drainage will in most cases convert a tract of land into a morass. Stagnant water is fatal to the life of useful vegetation, and it is here that the causes of the failure of many attempts to irrigate

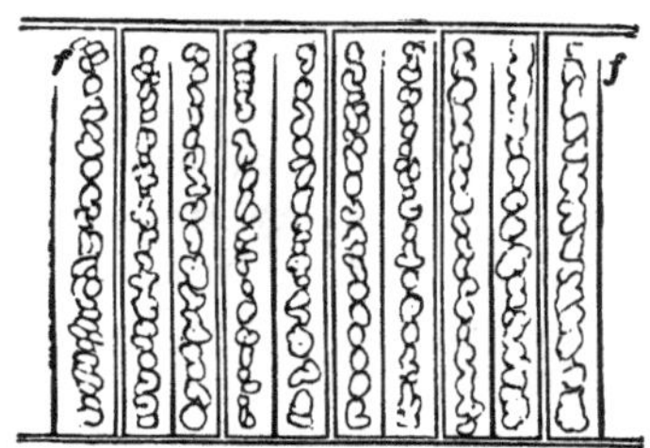

Fig. 40.—PLAN OF IRRIGATING A VINEYARD.

originate. In arid territories without rainfall, skillful irrigation will supply such a quantity as will be needed to supply evaporation from the surface of the soil and the transpirations of the plants. If more is given, the surplus must pass off through the subsoil, or remaining in it will work mischief to the crop. But such an excess of

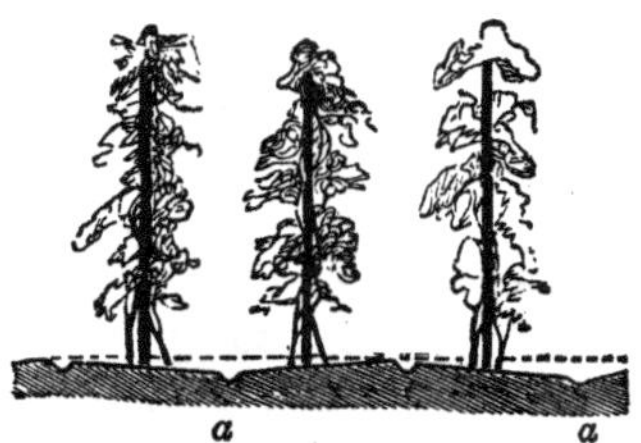

Fig. 41.—FURROWS AND DRAINS IN A VINEYARD.

water can rarely be procured in arid districts. On the contrary, the greatest economy must be exercised in using the limited supply, and waste is impossible.

It is otherwise in those parts of the country where partial or periodical irrigation is used. There the water supply may be copious, and the skill of the cultivator is to be exercised in conveying to his field only so much as may be serviceable and no more. But to hit the just mean is a matter of difficulty, if not impossibility, for several reasons. For safety, therefore, in these cases a system of drainage is imperatively needed. Especially is this the case in orchards and vineyards which are subject to so many varieties of blight and mildew, and other diseases which have their origin in atmospheric or meteorological conditions. Except in very rare cases, then, it will be imperative that a tile or other drain be laid in the subsoil at least four feet beneath the surface, between every two rows of distributing canals. This will remove the danger of injuring the plantation by excessive watering. The position of the drains is shown by the dark lines, *f*, *f*, *f*, in figs. 38 and 40, and by the small rings *a*, *a*, beneath the surface in figs. 39, 41.

The roots of trees seek out and follow a supply of water with great avidity. Drain pipes in orchards and gardens have been frequently penetrated by the roots of the trees and completely choked by a dense mass of fibers, eagerly appropriating the water found therein. For this reason the drainage of orchards by tiles is a somewhat hazardous business. To irrigate the soil of an orchard would tend to keep the roots near the surface where they would receive a sufficiently copious supply of water. With an abundant supply of water it is not probable that the roots would enter the drains, as the only purpose of their entrance there is to seek moisture. This being supplied as far as necessary upon the surface, the seeming instinct of the roots to enter and choke the drains would have no reason to exist, and would not be likely to occur. The great depth to which the roots of fruit trees and vines penetrate is undoubtedly due in part, if not wholly, to the effort to seek and procure sufficient moisture. The roots of vines have been found spreading at a depth of eight feet below the surface in soils that were naturally drained and not retentive of water. Although it is a matter of conjecture if the roots would descend so far when ample moisture may be found near the surface, the reasonable probability is that they would not. If the habit of deep growth should be a fixed one, it would be a question as to how deep the drains should be made in soils that are well supplied with plant-food in the subsoil, but were too retentive of water to permit a healthy growth at considerable depth. It is evident that with irrigation, and sufficiently deep drainage combined, the vine and fruit grower can render himself largely independent of seasons and locality, and give his vines and trees an ample depth of soil in which to spread their roots, and at the same time furnish them with all the moisture they may need near the surface. The practice will necessarily be modified by the character of the soil and situation; fruit growers, however, are rarely deficient in intelligence, skill, or patience, and are abundantly able to make such modifications of the general principles given in this work as may be needed. The practice in those countries where orchards and vineyards are irrigated is as follows: The periods of irrigation depend upon the heat of the season and the dryness of the soil. In the north of France and parts of Germany, water is given without any regularity, and only when the exceptional circumstances of the season make it needful. But further south, where the summers are hot and dry, and periodical drouths occur, fruit trees are irrigated constantly and vines periodically. The penalty for an excessive irrigation is a crop of fruit of inferior quality; watery, soft, and without flavor; the wood and leaf are pushed at the expense of the fruit; succulent fruits crack and burst, and shelled fruits have soft and imperfect husks. The effect of too copious irrigation upon nut-bearing trees is to develop the whole fruit simultaneously, the inner portions complete its growth while the woody husk is still soft, and the latter is either burst open prematurely, or fails to open at all, from want of the growing pressure of the kernels within. It is therefore necessary to act with extreme caution. Early fruiting trees require little or no irrigation, and late bearing ones are watered only after the fruit is set, and need to grow vigorously. As the ripening season approaches, the water is withdrawn, unless the necessity is absolute. During flowering no water is given at all, unless exceptional drouths occur, and then with moderation and at intervals.

The custom prevalent in the vineyards of the Crimea, a locality in Southern Europe, on the north shore of the Black Sea, and one subject to dry hot summers and cold bleak winters, is thus described by M. Clément-Bertron in the *Journal d'Agriculture Pratique*: "There are in the Crimea four valleys completely planted in vineyards to the extent of about 15,000 acres. The vines are irrigated each year as copiously as possible, not only during the winter, but from the termination of the vintages up to the season of the next flowering. Some growers even irrigate their vines after the flower is passed, but in general little water is given after the month of June up to October. As soon as the water has been applied, and the ground has dried, the vineyards are cultivated or dug over with the spade, and the vines are pruned. About 15 days before the vintage, the vines are clipped so as to give air to the fruit. After the grapes are ripe, there is no work done in the vineyard until the next season's labor begins. The cold of winter has not been found to injure the vines, although this is sometimes severe and long continued. The strength of the wines is not diminished by the process, the proportion of alcohol in them varying from 10 to 15 per cent. It is found that once the vines have been irrigated, the practice cannot be changed without loss of the product, and injury to the plants. Clear water is preferred to that which contains suspended matter."

The effect of irrigation is sometimes found to render both vines and trees subjected to it, very susceptible to the frosts and severe weather of winter. This disadvantage seems to be a necessary adjunct, or set-off, to the advantages gained by the practice. Thus, a severe winter has been known to destroy whole groves of olive trees that have been irrigated, while scattered trees, not so cultivated, have escaped. It is rare that we can altogether escape a combination of circumstances that seem to offer us only a choice of evils; an alternative, either side of which is about as disagreeable as the other; a Scylla and Charybdis, neither of which can easily be escaped; and this business of irrigation of fruit trees seems especially to be one in which the operator is obliged to exercise the greatest care and circumspection to avoid, on the one hand, the evils of excess, and on the other hand, the periodical and certain dangers which this practice enables him to obviate or mitigate when intelligently applied.

CHAPTER X.

THE IRRIGATION OF MEADOWS.

The permanent meadow is a very unusual adjunct to an American farm. Our climate is not naturally well adapted to the continued growth of grass. Our hot, dry summers are unfavorable. Generally it may be stated as beyond question, that the yield of grass is proportionate to the supply of water. As has been previously stated, no solid nutriment reaches any plant except as supplied to it in solution in water. What are the ultimate possibilities of growth in any crop is unknown to us, but it would seem as though they depended greatly upon the supply of water that can be absorbed, sufficient nutriment of course being provided. Rye grass, upon irrigated fields richly fertilized, has grown at the rate of one inch per day, and repeated cuttings have been made at intervals of 14 days, during a season of several months. Crops of grass upon irrigated fields of a total weight of more than **80** tons per acre, have been reported by trustworthy English farmers in one season.

Irrigated grass fields in Italy support easily two head of fattening cattle per acre, every year, and have long done so. In hundreds of localities in European countries are irrigated meadows, which have borne grass without any sign of deterioration within the memory of the inhabitants, or the knowledge of readers of local histories, although the crop has been cut and removed every year during this indefinite period. Whether or not these immense yields could be further increased by more skillful management is not necessary to inquire. These products are so far beyond the dreams of an American farmer, that they may well be considered fabulous. But there is no reason to doubt the facts. On the contrary, they should be used as a stimulus for us to adopt, wherever practicable, the methods by which these crops are produced.

The average product of grass upon our rich bottom lands, will not exceed two tons per acre, and upon uplands one ton per acre is a fair average yield. After a few years the best seeded of our meadows begin to deteriorate and run out. A change of crop is made and the meadows are once more seeded down to run out again in a few years. The cause of the failure is the heats and drouths which follow the hay harvest, and which cause a cessation of growth until they are past. Beneath a temperature which would be genial and invigorating to plant growth with sufficient moisture, the grass dies for want of the sustenance that water would afford. The most valuable crop we grow is thus reduced in its possible yield one-half or more. The only instance of an approach to permanent meadows in this country, is the few partially irrigated grass fields which are very sparsely located in hilly regions where springs and brooks are led upon the grass upon sloping hillsides. In these few cases, year after year, crops of two or three tons, and sometimes more, of hay are cut. Where a very imperfect irrigation has thus been employed for 30 or 40 years, the meadows exhibit no sign of deterioration. An occasional dressing of manure, and a little fresh seed now and then, keep them in a productive condition. But in the majority of these cases the water has been utilized for this purpose, from sheer necessity rather than from choice. A spring issuing from a

Fig. 42.—IRRIGATING A HILLSIDE.

hillside, or upon a level field, with high ground above it, and low ground below, either meanders wastefully through the level and escapes in an unsightly gulley at the edge of the hill, or it spreads over acres of ground, and makes a useless and unsightly bog. The careful farmer, to avoid this evil, and with an eye to thrift, leads the flow into a channel that departs slightly from the level, across the field and down the slope. A stone placed here and there in the channel, causes the water to overflow, and spread in a sheet upon the surface. One by one portions of the field are thus watered, and the effect is to induce a growth of grass that remains green beneath the snow, and grows luxuriantly as soon as it has disappeared, yielding two crops of hay in the year, besides some pasture when the springs cease to flow and the ground is capable of bearing cattle. Upon hundreds of farms in Pennsylvania, and in the valley of Virginia which has been settled by farmers from the former State, there are watered meadows of this character which yield a steady crop of hay, year after year, and possess a sod which promises to remain productive indefinitely, with its present treatment. This accidental use of the water has been in reality forced upon the farmer. Had it not been brought into a channel and confined to one or two canals, it would have flowed irregularly over the surface and have formed a morass. The process really has been one of drainage rather than of irrigation, and the reclamation of the surface rather than its studied improvement. The methods of watering meadows in common use are illustrated in fig. 42, in which a small stream is led down a slope, and at fig. 43, in which the stream is dammed and the water carried laterally as far as possible.

If such elementary and imperfect methods have been successful and profitable, how much more shall skillful and scientific irrigation add to the yield of our most valuable crop, and render possible the creation of permanent meadows, upon which grass may be grown in the greatest luxuriance, at an almost nominal expense! Numberless opportunities to make irrigated meadows present themselves everywhere. Far from being a matter of nicely arranged quantities of water, equally distributed at certain definite periods, as with other field crops; on the contrary, the irrigation of a meadow simply consists in causing a supply of water to pass over the grass at such periods as may be convenient; the convenience being only loosely circumscribed by times and seasons. It does not matter if the soil becomes saturated with water, it is only by the grossest negligence or ignorance that injury can be done. There is no danger, although the slope of the field may be considerable, of washing the soil, or cutting the surface into ruts or gullies. Water may be turned on to the sod without fear of excessive irrigation if it is only kept in motion. The more water that passes over the surface the more valuable nutriment is brought within the reach of the plant. Every blade of grass acts as a part of a filter which retains matter that may be either in solution or in suspension in the water which slowly finds its way over the surface. The mechanical resistance offered by the myriads of stems and leaves of the grass

Fig. 43.—IRRIGATING A RIVER-BOTTOM.

to a current of water are such that the combined effect is equal to a loss of head or level of 16 inches in 200 feet.

This retardation of the flow helps to cause the deposit of any solid matter suspended in the water, from which but few springs or streams are free, and also to bring every particle of the water into contact with the surface of the soil, or the surface roots of the plants. Not only, therefore, is the plant supplied with nutriment while the water is in contact with it, but a supply of nutriment is deposited and stored for future use. This freedom of application does not exist when cultivated or plowed lands are irrigated, and in their case more care and greater caution must be exercised to avoid injury. It is therefore advisable, in localities where only partial irrigation is needed, to cause these lands that are brought under the system, to bear grass in preference to any other crops, and to make the irrigation permanent and as perfect as possible. That is to say, that in all other than arid or rainless countries, meadows only, and no other field crops, should be irrigated, unless under exceptional circumstances; for the reason that the irrigation of a meadow is easy and requires but little practical skill, is more cheaply performed, because the works are permanent, and is more certain and profitable in its effects than that of other field crops.

It would not be difficult to give excellent reasons for these statements. It may be sufficient, however, to remark that, excepting in those districts where irrigation is needed for all crops, the water supply can rarely be made available for any other lands than river bottoms; for the reason that the coöperative effort of many proprietors would be necessary to bring a supply of water to a large tract, and this would be difficult or impossible to effect. Bottom lands are naturally suited to the growth of grass, and the means and the end of their irrigation match so accurately, naturally, and conveniently, that there seems to be a foregone necessity that the one should exist for the other. Further on, in considering more particularly the possibilities and methods of irrigating these lands, the advantages of keeping them as permanent grass lands will be still more conclusively shown if that need be.

Where the climate admits of it, irrigation of meadows is performed in Summer and in Winter. There are two objects in view. One is, to supply moisture to the soil at a season when there is an insufficient amount of rain, and the other is, to convey to the soil, and deposit upon it, whatever fertilizing solid matter the water may contain at a season when water is very plentiful. The first object is attained by Summer irrigation, and the second by irrigation in Winter. It is only, however, in those localities where frosts are neither severe nor long continued that Winter irrigation is admissible. Where light frosts alternate with sunny days, a covering of a few inches of water, gently flowing across the meadows, protects as well as fertilizes the grass. At this season the copious rains or melting snows carry into the streams an immense amount of fine, earthy matter, which may be arrested and caused to be deposited in a thin sheet upon the soil. In the course of several years this deposit has been known to raise the surface of the meadow many inches, every inch of this increase consisting of matter of the greatest fertilizing value. Where Winter irrigations can be made, they will be found of the greatest value, for they prepare the crop which is to be cut in the Summer by supplying in a great measure the necessary subsistence for its growth. Where the level of the field or the supply of water is such as to permit it, a constant current may be kept flowing over the surface during the period when growth is suspended, or from November or December until February or March.

Where it is necessary to make a series of levels to be irrigated in succession, each may in its turn be overflowed for a week ; or by arrangement of ditches and banks, the water from the upper level may pass over each lower one, supplying the whole, if it is in sufficient quantity. But where the supply of water is only limited, it is preferable to irrigate each level successively, for the reason that by far the largest quantity of suspended matter will be deposited by the first of the waters made to flow from one level to another, and in this case, the lower ones will receive a diminished quantity of deposit, in proportion to their distance from the source of supply. When the temperature falls sufficiently for ice to form, the quantity of water should be increased so as to keep a current constantly flowing beneath the ice. If the cold is sufficient to congeal the whole supply of water, so that ice rests upon the grass, the flow should be cut off. No injury will occur to the grass in this case, but if the water is still allowed to flow, the ice will be increased in thickness, and a longer time will be needed for it to thaw. If this imprisonment is continued too long, vegetation may be injured, but a week or two is insufficient to cause any injury.

In the Spring, when the water has been withdrawn and growth has commenced, there frequently occur cloudless nights and low temperatures, when hoar frosts are produced. On such occasions it is common to spread the water over the surface during the night, as a protection from the frost. The benefit derived is sufficient to repay the necessary care and labor during the months when these sudden changes are to be expected. There are many localities in the Middle and Southern States where this sort of irrigation might be practiced with very great profit. It is extensively practiced in Lombardy, where these "Winter meadows" are known as *marcite*, (*marcita* in the singular), and where they have long been known as the most productive of any meadows. As early as February, when the surrounding country may be yet covered with snow, these meadows, protected during the Winter by a covering of flowing water, begin to furnish their first cutting of grass. Five other cuttings follow, before the season closes, so that the cattle receive fresh grass during 11 months of the year. Twenty-eight tons of grass, or seven tons of hay, per acre, is the usual yield of these meadows. The valleys of several of the French, English, and Irish rivers, although subjected to a less genial climate than that of Italy, furnish many examples of successful Winter irrigations. Certainly a vast extent of the United States, where grass is a scarce product, might be made amenable to this profitable treatment.

At this point a typical case might be cited. When visiting England some years ago, the author's attention was attracted to some extensive water meadows upon the banks of a small river, the Mersey, which finds its exit into the sea at Liverpool. The upper part of this stream flows through broad, alluvial lands, which, before their reclamation, must have been marshy, and of little value. Extensive works have been in existence, however, for many years; precisely how long could not be ascertained by enquiry, all that could be learned was that "they were *always* there." The river banks were enclosed by dikes, or as they are termed on our Western rivers, "levees," sufficiently high to prevent overflow, even in freshets. Substantial water-gates were made in these banks, leading into lateral channels at right angles to the river. These lateral channels had banks of equal hight and solidity with the main banks. The lateral banks extended from the river until they reached the gradually rising ground at their level. From these, other banks, enclosing lesser canals, with water gates at their heads, and parallel with the river, extended until they met the next range of cross banks; thus dividing the broad bottom lands into a series of parallelograms enclosed in a system of canals at right angles to each other. From these canals, gates sliding in perpendicular grooves, and raised or depressed by racks and pinions, opened into the meadows. When the level of the river was raised by unusual rains, the gates were opened, and the meadows enclosed within the different canals were flooded with water, to a depth of about six inches. So long as the river remained high, the gates were opened sufficiently to permit a gentle flow of water from one section of meadow to another, until it escaped into the river again at a lower level, by drains through the banks; or the water remained upon the meadows, in a state of quiescence, to deposit upon them the fertilizing matter which it held in suspension. For centuries this practice had been followed, and the grass thus grown had been mowed and fed to cattle, or made into hay. The same practice was afterwards observed in other parts of England, Ireland, and in Continental Europe, where scarcely a possibility of utilizing a stream in this manner has been neglected.

What is there in our circumstances that prevents the practice of so great an economy ? There is no reason why our thousands of rivers might not each have its scores of watered meadows, along its banks. The skill to execute the necessary work is abundant. Hundreds of civil engineers, relieved from duty upon the suspended or finished railroads, might profitably turn their attention to this branch of their profession, if only farmers were alive to the advantages of thus improving their farms.

The system adopted in Europe may be applied here with the greatest facility, but upon a much larger scale, as our rivers are larger, and our river bottoms more extensive. The irregular and unrestricted, and therefore sometimes destructive overflows would thus be controlled and profitably utilized. The supply of grass, our most valuable fodder, would be greatly increased, and a needed improvement would be effected in our agriculture.

In the Northern States and Canada, Winter irrigation is impracticable, and there Summer irrigation only would be beneficial. As soon as the ground is free from frost, the water of the streams, highly charged with sediment, might begin to be utilized. Afterwards, when growth has begun, no check would be permitted, but every night during a dry season, the meadow might be flooded. Then, when the crop, brought to an early maturity by the stimulus of abundant moisture should be cut and removed, a new growth would be forced, and under the influence of a genial sun, would advance quickly. Two crops could be made by August, and in many cases a third could be procured by October. The economy of the system is sufficient to permit a considerable outlay in preparing the surface, and in addition there might be estimated a vast saving by the substitution of growing grass to be cut and fed to cattle, for the present costly practice of pasturing. Nevertheless it is not necessary that pasturing be abandoned, for irrigation is as applicable, to a large extent, to pastures as to meadows.

The details of the methods here alluded to will be treated of in a succeeding chapter.

CHAPTER XI.

THE USE OF SPRINGS FOR IRRIGATION.

Springs are one of the sources from which water for irrigating meadows is most frequently procured. They are often situated advantageously, so that the water may be circulated by gravity over the land on a lower level. It is possible in many cases to reach the actual source of the spring at a point several feet above that at which it naturally issues from the ground. A vast number of springs really furnish a much larger supply of water than is suspected. Usually they are allowed to saturate the surface and escape into the subsoil by numerous hidden channels, which in the aggregate would furnish a respectable stream. By proper economy in using the water, a very small stream may be made to irrigate a field of considerable extent. It is by using water in driblets that many springs are wasted. A stream yielding one quart per second may have its water wholly swallowed by the thirsty soil within 200 feet of its source, when by arresting the flow and accumulating it in a reservoir, which may be discharged at intervals by automatic arrangements, the water may be made to escape in a volume four times as large, and sufficient to cover eight times the surface.

By this contrivance a very small spring may be utilized. One yielding 2 quarts per second will serve to water four acres of meadow if stored for 24 hours, and discharged periodically at intervals of that length of time. During this period 43,200 gallons would be accumulated, which would supply nearly one quart of water to every square foot upon the four acres; a very ample allowance in addition to what is furnished by the rainfall, to secure a full crop of grass. It would be preferable to accumulate a larger quantity of water than this, if possible, and to give a more copious watering less frequently. A thorough saturation of the soil at intervals, as has been before explained, is better than more moderate waterings more frequently given. Air is as vital a necessity to vegetation as water, and if access of air is denied, the roots of the plants must perish. Where water goes, air follows, and as evaporation takes place, air fills the space previously occupied by the water. To moisten the soil to a depth of several inches gives that coolness which the grass roots find necessary for their healthful growth; but to moisten the soil to a depth of only an inch or two, gives no supply sufficient to resist the drying effects of the sun's heat, or a hot dry summer breeze. Two inches of water given every week would be a very good supply, and with a spring of the size of flow mentioned, economically stored, twelve acres of grass could be watered once a week. The effect would be equivalent to that of the fall of a steady, moderate shower during a whole day and night, and occurring every week, and every farmer can readily understand the value of such a shower upon his meadows.

To store 43,200 gallons of water will require a reservoir of 5,760 cubic feet. One 40 by 20 feet, and 7 feet deep, will have about this capacity. If the width is doubled, the depth may be decreased one half. The shallower it it can be made the better for many reasons. The temperature of spring water is generally too low in the Summer for immediate use, and its value is greatly enhanced by being raised to an equal or greater temperature than that of the air. This is most quickly done by exposure in a shallow pond. Every foot saved in depth is a foot added to the level of the outlet, and so much more added to the area that may be irrigated. This is evident, because if the reservoir is 7 feet deep, the surface of the water can be no higher than the level of the source, unless the water is pumped up into the reservoir, and it is clear that the water discharged cannot be made to irrigate any land that lies higher than the bottom of the reservoir. With a 7 foot reservoir, all the land that lies between the levels of the bottom of the reservoir and the surface of the water cannot be irrigated ; unless there are

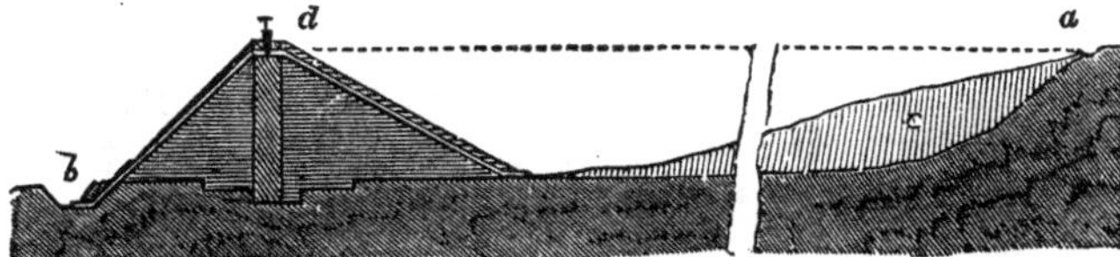

Fig. 44.—SECTION OF RESERVOIR.

several discharging pipes at different portions of the reservoir. With regard to cheapness of construction, if not to effectiveness in operation, it will be found far better to have the reservoir as large as possible, at least of sufficient capacity to contain water enough for use every two to seven days.

Where the surface slopes but one way, an embankment may be made on three sides of a square, inclosing a sufficient space, and open on the upper side at which the spring will discharge itself. This is shown at fig. 44 in section, and in plan at fig. 45. To irrigate the strip of land parallel with the reservoir, a canal or furrow may be carried on a level with the spring, seen at *a*, *a*, in the figures, to the boundary of the meadow. The overflow from the reservoir may be made to pass into this canal.

This will be found a very convenient arrangement. Figures 44 and 45 are intended to represent a typical form of such a reservoir as this. The spring, escaping by a small stream, seen in the plan, fig. 45, occupies the

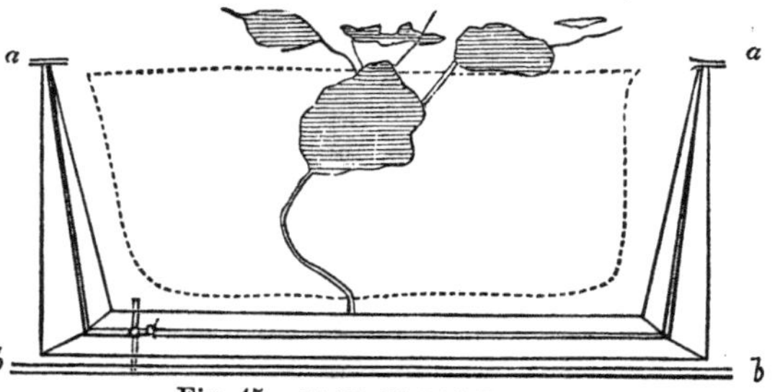

Fig. 45.—PLAN OF RESERVOIR.

point *a*, in fig. 44. The ground around and below the spring is excavated as shown by the dotted line, (fig. 45), and by the part lightly shaded, marked *c*, in fig. 44. The earth removed serves to make the dam which is construct-

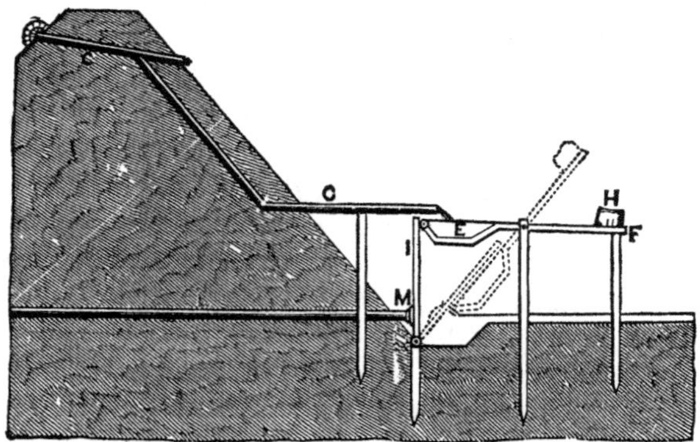

Fig. 46.—TRAP FOR DISCHARGING RESERVOIR.

ed in the manner hereafter described. (Page 111). A pipe is laid in the dam, for convenience not far from the surface, and a valve, operated by a key, *d*, closes and opens the pipe. The pipe is in fact a siphon, and if opened when the reservoir is full will discharge until the water is exhausted, into the distributing furrow, *b*, fig. 44, and *b*, *b*, fig. 45. The dotted line, in fig. 44, shows the level of the water in the reservoir when it is full and overflowing at the outlet, *a*.

When the reservoir is filled, the surplus is discharged on each or either side, by the channels made for that purpose. This will obviate the difficulty previously pointed out. The flow may then be turned upon the upper portion of the meadow for twelve hours, in such a manner that the whole of the water shall be absorbed by the soil, and afterwards the contents of the reservoir may be flowed on to the lower portion during the next twelve hours, when the outlet will be closed. Many different arrangements for the use of the water may be devised to meet the necessities of any peculiar case, and as experience is gained, any difficulty that may arise at the first will be readily overcome. The reservoir may be discharged by an intermittent self-acting arrangement which is either a siphon, already described, or a more complicated but equally effective method of a balanced trap, fig. 46. The balanced trap consists of a board having a weight, *H*, attached to one end, and a cup or basin at the other, and being suspended upon pivots in a frame erected at the edge of the main distributing ditch at the outlet in front of the dam. The board is nicely balanced, so that when the basin is empty the weighted end rests upon a prop, *F*, purposely placed for it; but when the basin is filled with water it overbalances the weight and falls. As it falls it releases a gate, *I*, upon which is fixed a leather cushion which closes the outlet pipe of the reservoir, *M*. When the reservoir is empty the gate is raised and the pipe is closed. When the reservoir is filled the overflow enters a pipe through the upper part of the dam, *C*, and flows into the basin. The basin descends and releases the gate; the force of the water flowing from the discharge pipe keeps it open as long as the stream is running into the canal. When the water is exhausted the pipe is again closed. To prevent the water flowing over the dam, through any accidental stoppage in the machinery, a branch of the overflow pipe is carried down the face of the dam into the canal. This apparatus is of very general use in the Swiss Cantons, and in irrigating works elsewhere, and works with regularity and precision. It is necessary that the balance-trap be properly adjusted and looked after occasionally. The worst that can happen in case of accident is the overflow of the reservoir by the pipe into the canal without harm. If the overflow is provided for at the inlet by a pipe or a channel placed there, as already suggested, this overflow pipe in the dam will not be needed. In practice, however, it will be found safest to have every guard against accident and consequent damage to the works, and two outlets will be twice as safe as one. If it is thought desirable, the waste-pipe in the dam may be placed two inches above the level of the other outlet, so that it will come into use only in case of a stoppage of the lower one. The outlet pipe should be large enough to discharge the water as least four times as rapidly as it enters the reservoir ; so that the storage of two days flow may be discharged in a night or during one cloudy day. (Under no circumstance should the water be permitted to escape during the day when the sun is shining). A three-inch pipe will discharge nine quarts per second, which would be more than enough to furnish two inches of water to four acres in 12 hours. A pipe of this diameter would therefore be of ample size for a 12-acre meadow, giving a weekly watering to each 4 acres by three discharges of the reservoir.

A siphon is not always to be depended upon to discharge a reservoir automatically. Sometimes the water, when rising slowly and not filling the pipe completely, trickles over and does not set the siphon in operation.

When an arrangement is made for the safe overflow of the surplus in some manner, a valve may be attached to the head of the siphon, (*d*, fig. 44,) by which the flow may be started, or a tap may be fixed to the lower end of the pipe for the same purpose. This would be preferable to the plain siphon, although it would involve the necessity of personal attendance at stated times to discharge the reservoir. But no one should undertake the irrigation of land who is averse to giving the necessary attention to the details at proper times. An unexpected accident, the work of vermin, the presence of some floating body, or some other trifle, may stop the work, and unless some oversight is given to it, mischief and loss might occur. It is therefore advisable to depend upon personal effort rather than automatic contrivances, although it may be as well to have the latter in use if it is not made an excuse for neglecting careful supervision. Of all automatic arrangements for discharging the water, the balanced trap is the most trustworthy one.

Where the surface is not regularly sloping, a hollow or ravine may be made into a pond or reservoir by building a dam across the hollow. In building any dam of this character, the foundation must first be excavated until the solid subsoil is reached, or the dam will leak and its stability be destroyed. A trench at least a fourth of the width of the dam should be dug and filled with puddled earth or clay. The front and rear of the dam may be made of sods cut from the bottom of the reservoir, and the center up to the top should be made of earth or clay puddled and rammed solidly between the walls of sods. The dam, if a high one, should be at least twice as wide at the bottom as it is high ; and the width of the top should be one-fifth that of the bottom. The inner slope should be 18 inches horizontal to one foot of hight. The bottom of the pond should be made of puddled clay to prevent a waste of water. A section of the dam is seen at fig. 44; the hight of which is 8 feet, width 16 feet, and the puddled clay wall in the center is shown by the darkly shaded portion. Where the spring is of sufficient volume to supply all the water that may be needed, it would still be worth while to provide the reservoir for the sake of gaining the increased temperature; but in such cases the reservoir will not be needed for the purpose of distribution, but only to warm the

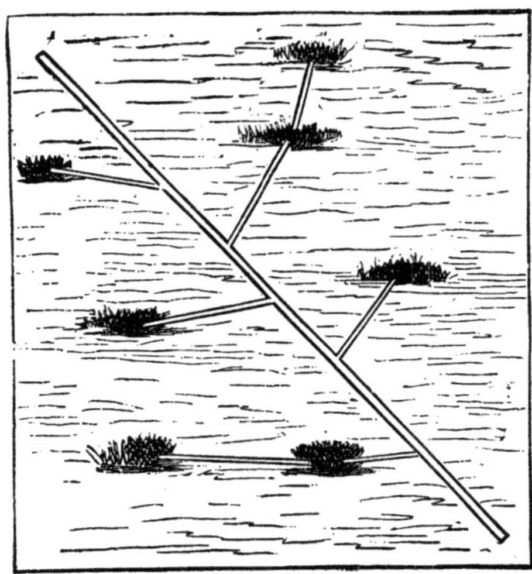

Fig. 47.—MANNER OF COLLECTING THE WATER OF SPRINGS.

water. The overflow, then, only will be used, which will escape on the same level as that of the inlet. The course of the current through the reservoir should be made as circuitous as possible by means of a division of boards in the center, that the exposure of the cold water to the warm air or sun's heat may be the longer. When water is retained solely for this purpose, the space in which it is confined should be large and shallow, so that the exposure of the water to the sun's heat, and the influence of the atmosphere, may be as thorough as possible.

The temperature of the water has a considerable effect upon the growth of grass. Every one has noticed the effect of a warm shower, in early Spring, in starting vegetation; and also the ill effect of a cold rain, in the Fall, in arresting growth. In all cases the water should at least be of an equal temperature with the air. When spring water is used, the temperature can only be raised by exposing it in ponds or reservoirs for a time, and the shallower the pond the more quickly will the water be warmed. Exposure to the atmosphere also exerts a chemical effect, and some waters that contain sulphate of iron, or other deleterious substances, are rendered harmless by the oxidation of these impurities. Thus the temporary storage is of sufficient advantage both in enabling an intermittent irrigation, and in warming and purifying the spring water, to make the cost of the reservoir and distributing apparatus a profitable expenditure for any meadow of not less than four acres in extent.

It is often the case that a number of springs exist upon the surface that may be brought together into one channel with great economy. A spring is often merely the overflow of underground streams, and by digging downwards the whole of the water may be captured and brought into one channel, with the double advantage of draining a wet field and of utilizing the water for the irrigation of a meadow below the level of it. The diagram, fig. 47, represents a case of this character. A number of springs break out at the surface, and spreading make a marsh, but form no stream there. To utilize the water of these springs, and to drain the wet surface, all that is needed is to

Fig. 48.—THE DRAIN AND DISCHARGE PIPE.

cut a drain (see fig. 48) from each of them, leading to a common channel, and deep enough to reach the subterranean sources from whence the overflow comes. The main channel is made to discharge at a point required either into a cistern or into an irrigating ditch. The method of making the drains need not be costly. If stone is at hand, and flat long pieces can be easily procured, the drains may be made by placing long narrow stones against the sides of the ditch, at the bottom, and covering them with shorter pieces placed crosswise. Small fragments may be thrown upon these and earth upon them. This is shown at fig. 49. If round stones only can be procured, the drain may be made as shown in figs. 50 and 51. The depth of the drain, should not be more than is necessary to reach the main stream, as for every foot deeper than that, so much head at the outlet is lost, and so much less land can be watered. In digging the drains, for the

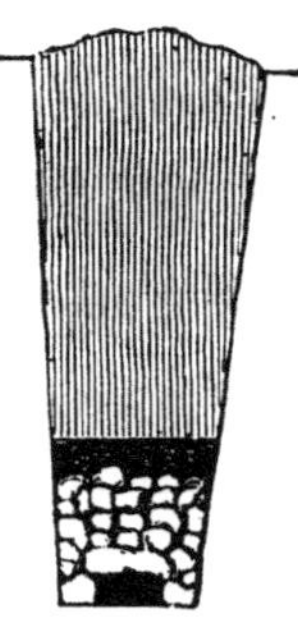

Fig. 49.
FLAT STONE DRAIN.

Fig. 50. ROUND STONE DRAINS. Fig. 51.

same reason, no greater fall should be given than is needed. Six inches in 100 feet is ample fall to keep the drains clear from sediment, and more would probably result in washing out portions of the drains at the sides or bottoms. A very useful level for laying out the drains may be made as shown at fig. 52. It consists of a parallel-edged board, seven or eight feet long, with a ⊥ affixed near one end, which supports a pendulum. A scale is marked on the board at the foot of the pendulum, whereby its motions are noted. When the board is perfectly level the foot of the pendulum marks 0. When the board inclines either way it varies accordingly. A handle is fixed to the end of the level, which serves to hold it in position when in use. In case it is not wished to lay out

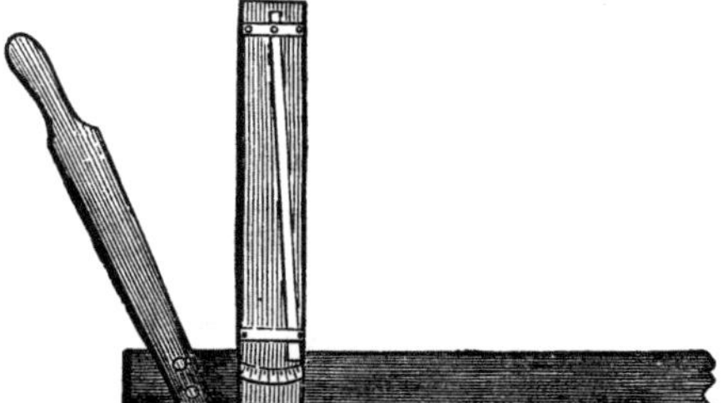

Fig. 52.—LEVEL.

the bottom of a ditch to a very accurate grade, the mere movement of the pendulum to the right, when looking at the scale or index, will show that the grade is downwards. But if accurate measurement is desired, it will be necessary to make the instrument in proportion, and mark the index carefully also with a proportionate scale. Thus, if the bottom of the level is six feet long, and the ⊥ two feet high, an elevation of the hinder end of the instrument of half an inch would be equal to a grade of one inch in 12 feet, or one in 144, or eight inches in 100 feet, and would cause a deviation from the perpendicular of the pendulum of one-sixth of an inch; a grade of 16 inches in 100 feet would cause a deviation of one-third of an inch. If such close measurement is desired, the instrument will have to be carefully made. For ordinary operations, it will only be necessary to take care that the ⊥ is set on quite square, and then the least movement forward of the pendulum will show the grade to be correct.

When the waters of springs, such as are now under consideration, are to be used directly in irrigation, the method shown at fig. 53 may be applied. The springs *s*, *s*, *s*, may be opened or cleared of rubbish, and may be led directly into furrows following the lines of level shown by the dotted lines. Or they may be led into the larger springs and the collected water be discharged as shown at *S*, *S*. Or several springs near the center may be

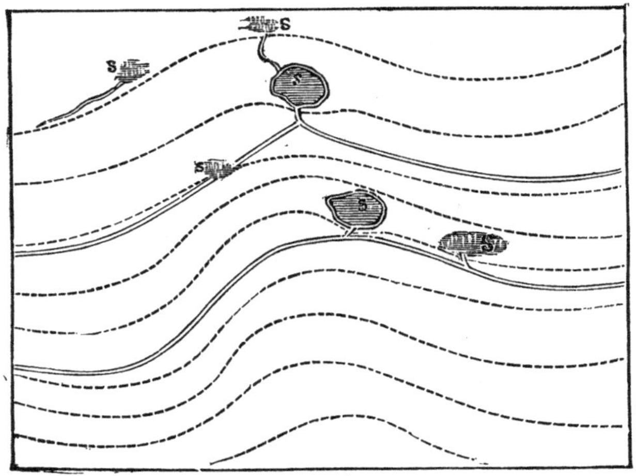

Fig. 53.—DIRECT USE OF SPRINGS.

gathered into a pool or reservoir, and the others led into it, and the whole supply be discharged into a main furrow following the level as seen at fig. 54, in which the springs are seen at *s*, *s*, the reservoir at *R*, and the irrigating channels at *c*, *c*, *c*.

By this management the drainage of wet, arable lands, also may be made to furnish a supply of water to irrigate meadows, and the instances where such a combination of advantages may be availed of are far from scarce or few. Indeed the swamps that now produce very inferior herbage, or that are totally useless, or worse, because productive of miasma, or dangerous to cattle that may trespass upon them, and that might be reclaimed by drainage, and at the same time furnish a copious supply of water for irrigation, are far more numerous than would be suspected by any but an engineer, whose practiced eye can see at a glance the possibilities in this respect that others would fail to perceive. It nine cases out of ten, at least, a swamp is in reality a spring, or a number of them, which spread

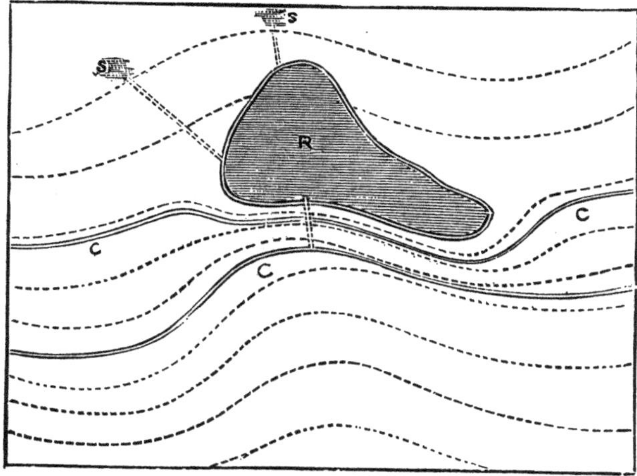

Fig. 54.—THE SPRINGS COLLECTED INTO A POOL.

themselves over the surface and stagnate, losing their flow by evaporation or slow filtration through the surrounding soil, or their own subsoil. To utilize this waste water would be to turn a diseased and pestilential spot into a healthful and productive field, that would also contribute the means of enhancing the productive capacity of neighboring fields. Then "out of the eater cometh forth meat," and out of the waste place cometh forth fertility.

CHAPTER XII.

FORMATION OF WATER MEADOWS.

Every American farmer will acknowledge that grass is the most desirable, but at the same time the most difficult crop he can raise. It costs less to raise than any other crop when the adverse climate can be vanquished. But fortunately the American climate it not invincible, and there are means by which this crop, (as well as others), may be cultivated with success, in spite of heat and drouths. One of these is the system of irrigated, or water meadows, upon which the growth of grass can be made continuous during both Summer and Winter, for where the climate is not sufficiently cold to form ice more than two inches in thickness, grass may be kept in a growing state throughout the Winter, and be made ready for the first cutting in February or March. The United States is the only civilized country in which grass is not so grown, more or less. There is scarcely a river in Europe whose waters are not compelled to nourish and protect thousands of acres of its bottom lands wherever they can be brought upon them by means of embankments and ditches. On every hand the observant traveler sees irrigation works of extensive and substantial character, and of great antiquity ; and verdant meadows within them, covered with the most luxuriant vegetation. These works are to be found where the climate is naturally as unfavorable to the growth of grass as in any of our Southern States, although it is true that in warm, humid climates, or those where the heats of Summer are not so ardent, water meadows find their greatest developement. The small county of Wiltshire, in England, alone has 20,000 acres of water meadows, most of which have been in cultivation for over 150 years. This county is a famed dairy county, and the Wiltshire cheese is a staple product in the markets of the country.

But it is drouth, and not heat alone, that is fatal to the growth of grass, and which sears it as the breath of a furnace. Heat and moisture develop vegetable growth most abundantly. Without declaring that irrigation is to revolutionize our husbandry, it is only necessary to refer to the abundant opportunities which exist here for enterprise in this direction, to be assured that a vast change for the better would occur if it were brought into general use. It is a mistake to suppose that an irrigated meadow depends solely upon the use of water during the Summer months. On the contrary, wherever it is possible to be done, it is by application of water during the Winter season, or from the Autumn to the Spring, that the crop gains an accumulation of strength which enables it to pass through the Summer in safety, giving several crops in that season. Not that Summer irrigations are not useful or necessary, but that they are of less volume and of less continuance.

The chief advantage of this system is the accumulation of fertility made during a period when otherwise the ground is wasted by rains, and denuded of soil and soluble matter that it is not in a condition to spare. The meadow is made the place of deposit for a large portion of the matter of which other lands, not so improved, are deprived by rains and floods, and if the whole of the waters of the streams could be arrested and made to give up their burden, the whole of the value lost by them would be regained, and none escape to the sea or the estuaries of the rivers to form future lands of the richest character. The opportunities for producing grass upon water meadows in the Southern States, where Winter irrigation is possible, and where the river flats are extensive and numerous, are many and great, and the advantages in this direction are too important to be neglected.

The nature of the herbage upon an irrigated meadow depends greatly upon the skill with which the irrigation is managed. If water is used in excess, the more valuable grasses disappear and inferior ones take their place, such as quack grass (*Triticum repens*), the spear grasses (*Glyceria aquatica*), and *G. fluitans* and other coarse species. By careful management, re-seeding, and manuring, timothy and clover may be retained in a watered meadow, but there are several grasses which are but slightly inferior to timothy, and which grow abundantly and constantly, that are much better adapted to this culture. These are the fowl meadow grass (*Poa serotina*), rough-stalked meadow grass (*Poa trivialis*), the tall meadow oat-grass, called ray grass in France, (*Arrenatherum avenaceum*), and the well-known red-top (*Agrostis vulgaris*).

These grasses furnish a heavy burden of sweet, nutritious, palatable hay, and immediately after mowing, when watered, spring into a vigorous new growth. Italian rye-grass (*Lolium Italicum*), is extensively grown upon irrigated meadows in England, and yields repeated heavy cuttings of forage for soiling. It has been tried here without success, but not on irrigated lands. It is probable that under irrigation it will be found of equal value to other grasses that have already been naturalized, and are known to be available, as it is the chief grass grown upon the Italian water meadows, upon which it yields several cuttings, equal in the aggregate to 30 or 40 tons of green fodder per acre yearly. A mixture of five to seven pounds each of the four varieties named, as best adapted to watered meadows, would give a thick growth, and as some of them increase from the roots, a thick permanent sod would be formed, which would be in active and successive growth up to October, or even later in the season.

The undulating character of the surface of the soil offers the greatest facilities for using the waters from streams, both small and great, in irrigation. There are millions of acres upon the banks of streams that could be made to bear crops of grass permanently, with the greatest profit, at a comparatively small outlay per acre. It is where the surface to be irrigated is large that the process of irrigation is the cheapest. Where a stream flows naturally above the surface of a portion of the neighboring land, the cost of irrigating the land will be very small, and the cost per acre will be the minimum when the supply of water is abundant and the area to be watered is large. In this case no dam will be needed, or at most such a one as can be made at a small expense and maintained with little trouble. A simple barrier of stones, or a few planks, or a log laid across the stream and held in its place by a few stakes driven in the ground, will suffice to divert the flow into a canal, which will lead the water with the least possible loss of level to the ground to be irrigated. A narrow valley having a stream meandering through its center, and with sides gently sloping toward the stream, is peculiarly well adapted for irrigation. The whole length of the valley, from its head to its outlet, may be made a succession of meadows. The small tributary streams of the valley will be made to aid in the work and contribute their share to the general supply of water.

Should the streams be subject to early Spring and late Fall freshets, so much the more valuable they will be. Every flood will bring down a large amount of solid matter to be deposited as a fertilizer upon the soil. The water of floods is also highly charged with soluble matters which are rendered up to the soil through which it is made to percolate. The only disadvantage is, that should a flood occur when the grass is nearly ready for cutting, a considerable quantity of sand may be deposited upon it, and much of the crop may be lodged. But this difficulty is unavoidable, and would occur in any case, and must be submitted to as one of the drawbacks incident to the

6

process of doubling or trebling the usual amount of the crop.

The first business to be undertaken in forming such a meadow is to thoroughly drain the land either by underdrains or by open drains. The most important drain will be that which cuts off all the springs which issue from the foot of the uplands, and which generally render the low land a sodden marsh. Frequently this drain should be dug to the depth of six feet, that every spring that may issue below may be intercepted and tapped. This drain should be cut above the highest level to which the irrigating ditch can be carried, and may discharge into it or be carried beneath it and made to issue in the lateral drains. Next, the surface is to be leveled, the hillocks cut down and the hollows filled, so that no stagnant water can be retained in them, and the lateral slope of the meadow be made perfect up to the edge of the stream. The stream, or so much of it as can be used, is then diverted into side channels, which are carried as nearly upon a level as possible until they reach the foot of the upland, when they are carried still upon a level or with a slope of not more than one foot in a thousand, in a direction parallel with the general course of the valley, but yet following the winding made necessary by the configuration of the surface. The general arrangement of the dam, canals, and drains, is as follows : see fig. 55. The winding stream which occupies the center of the valley, shown by the dotted lines is straightened, and dammed at *a ;* the lateral canals are carried each way from the dam to the borders of the valley, and from them a regular system of distributing canals is supplied. The main cross drains, *b, b*, are above the canals on either side, and the drains, shown by the dotted lines, are carried directly to the stream, or they may be made to discharge into the water furrows if so desired. The level of the stream may be raised by embanking its sides for a sufficient distance, instead of building a dam across it and forming a pond. But the value of a pond upon a farm, if for no other purpose than procuring a supply of ice, would amply repay the value of the land and labor in one year. The

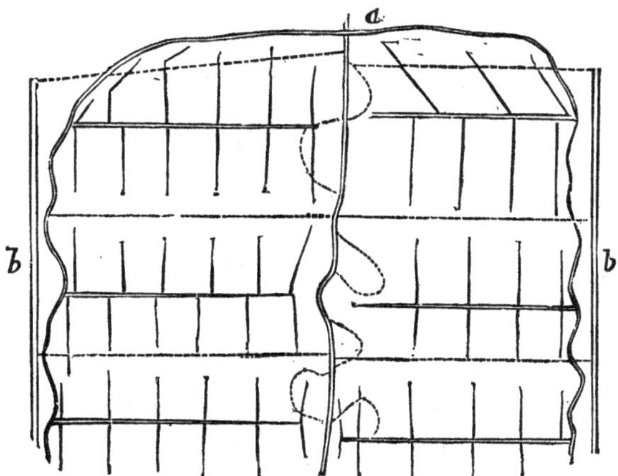

Fig. 55.—IRRIGATION OF A VALLEY.

arrangement of canals here described is a typical one for this kind of meadows; it is capable, however, of abundant modifications, to suit varying circumstances. It is given to illustrate the principle upon which these meadows may be formed.

There are various other methods of raising the water than this which has been described, some of which may be mentioned as being applicable to various circumstances. The old-fashioned *noria*, which has been in use in Southern and Central Europe since the eleventh century, is not yet out of date. It is still used in Savoy, Lombardy, Spain, and parts of France, and being easily constructed, and cheaply effective, where the supply of water is sufficient, might be used in some cases here. A wheel, having broad floats, is hung upon an axle, so that the lower floats are submerged in the stream, fig. 56. By offering a little obstruction to the stream, to increase the rapidity of the current where the natural velocity is not sufficient, the wheel is set in motion and revolved. Water buckets are fixed to the circumference of the wheel, in such a position that the direction of their longitudinal axle is 45 degrees from that of the axle of the wheel. The buckets are

Fig. 56.—THE "NORIA" OR WATER WHEEL.

partly filled as they pass through the water, and are discharged as the wheel brings them round to an inverted position, into a wooden trough placed alongside of the wheel. From this trough the water is conveyed to the distributing channels. Water may be raised by this rough and ready process, in the cheapest manner, to a hight of ten or twelve feet, requiring no attention and working by day and night so long as the stream flows. Another method by which a small portion of the water may be raised is applicable to brooks of moderately small volume, as well as larger streams, viz., the use of a water-wheel. Where the stream cannot be raised conveniently, an undershot wheel may be set in motion by turning the current into a wooden trough or shute, and impelling it against the floats of the wheel. Where a dam can be made, an overshot wheel may be used. Either of these wheels may be made to operate a chain pump, and raise a considerable amount of water. This pump is preferable to any other, as there are no valves to be choked by small floating substances, or to be worn by sand, which may be brought down by the stream. Wooden pins may be inserted around the rim of the wheel, from which a wooden pinion or gear may convey the motion by a short shaft to the pump.

The most economical form of meadow is the "water meadow," which is one so arranged that it can be flooded completely to a depth of several inches, and the water can either be retained upon the surface when desired, or made to pass over it with a slow, steady current. These are the meadows which in parts of Europe are so productive of grass, being protected during the winter from the slight frosts or snow which would stop the growth of the herbage, by a covering of water. Where the land cannot thus be completely covered, meadows cannot be irrigated in the winter season, in climates subjected to frosts sufficiently severe to freeze the ground an inch in depth. The too well known destructive effects of a frost upon a sod saturated with water, entirely forbid Winter irrigation in the Northern States. But in the Southern States, where frosts do not continue more than a few days at a time, the "water meadow" may be made a valuable addition to the farm, and supply such an increased amount of fodder for stock as may easily change the system of farming to a very considerable extent.

In forming water meadows no dams are used, nor is any water raised above its level. The streams are embanked so as to confine the water which is diverted from them and is carried in a level channel which gradually diverges more and more from the stream, until the whole of the land to be brought under treatment is inclosed. As the level of the surface slowly descends, that of the canal

Fig. 57.—SECTIONAL PLAN OF WATER MEADOW.

rises gradually above it until there is a difference of at least a foot between the levels of the water and the ground, at the upper portion of the meadow. The more regular the slope of the meadow the better in every way. If a perfectly smooth surface can be made, the meadow is then a perfect one. A perfectly formed meadow is the one that lies in a succession of smooth, gently sloping

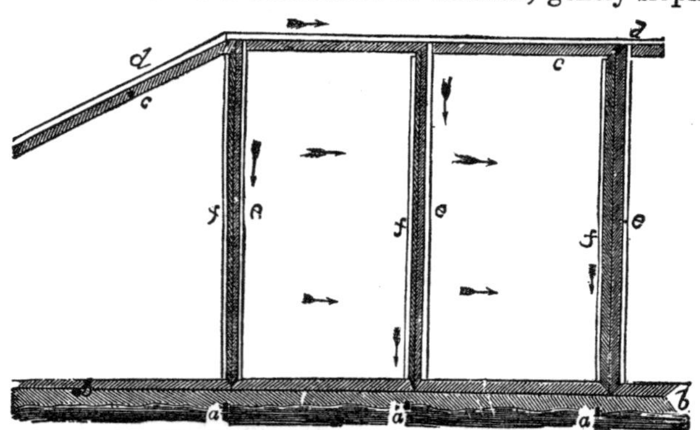

Fig. 58.—GROUND PLAN OF MEADOW.

tables, each one one or two feet, or more, below the level of the other. A meadow so prepared will show a section similar to that in fig. 57, in which the irrigating canals are seen at *e*, *e*, and the collecting drains at *f*, *f*. Spouts in the banks, at *a*, *a*, may pass the water from one level to another. (See also page 113.)

Each portion of the meadow will be confined between banks upon the sides, one of which will be upon the edge of the river, and the other upon the opposite boundary, which is the main supply canal, and between a canal of distribution at the head and an open drain at the foot.

Fig. 59.—SELF-ACTING WATER GATE.

This is shown in fig. 58, in which *a, a,* is the river; *b, b,* the river bank; *c, c,* the opposite bank; *d, d,* the supply canal; *e, e,* the distributing canal; and *f, f,* the drain. The drain discharges into the river through the bank by a self-acting gate, (fig. 59,) which yields to the outflow, but is closed by an inflow from the river. Or the surplus water from the upper level may be discharged into the distributing canal of the next lower level. The water is passed from the supply canals to those of distribution, either by a gate raised by a winch and pinion and rack, fig. 60, or a spout through the bank of the canal, which is closed by a slide, seen in fig. 61, and at *a, a,* in fig. 58.

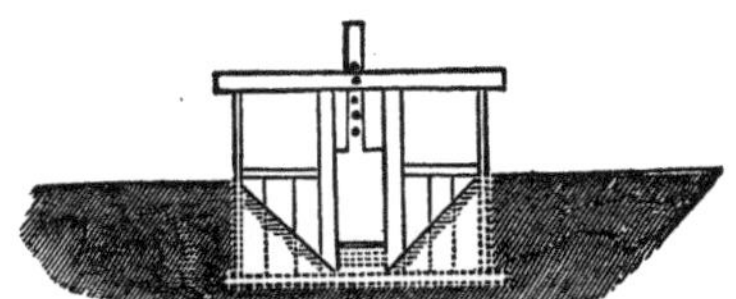

Fig. 60.—WATER GATE.

The water from the canal is first turned upon the upper level; when this is covered to a proper depth the gate is closed, and the water turned through the next gate upon the next level, and so on until all are covered. A sufficient quantity of water is allowed to pass on to each level to maintain the proper depth, and allow a gentle current to flow from the drains. This is important when the temperature falls below the freezing point. Observations have been made, which have shown that when this has occurred, and the temperature of the air has been as low

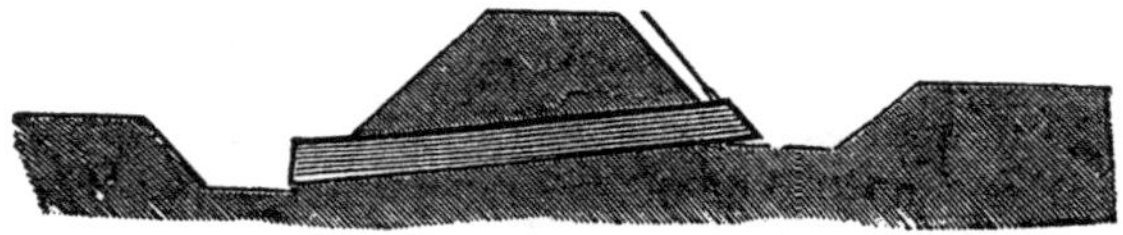

Fig. 61.—SPOUT IN THE BANK.

as 26°, that of the grass beneath the ice has been no lower than 42°, and that vegetation was still active, as shown by the color of the verdure.

As regards the amount of water used, and the manner of using it, the following experiences may be cited.

A comparison of fields that have been less abundantly watered, with those that have received a copious supply, has shown that the crops upon the latter have been infallibly increased.

Where during one Winter the irrigation has been suspended, the succeeding crop has been little or nothing.

Where the water that has passed over a field has been flowed upon another, the crop of the latter has been very inferior to that of the former, showing conclusively that the earth had completely abstracted the fertilizing property of the water in its first contact with it.

In proportion to the abundance of water supplied during the Winter, so is the yield of grass in the Summer. In short, facts are conclusive to show that the quantity of water that can be used, is the gauge of the harvest to be expected. The Winter irrigation supplies the fertility, that of the Summer simply supplies the necessary moisture. In this respect the action of water constantly passing in a thin sheet over a grassy sod has a different effect from that of water passing over uncultivated soil. It does not wash the soil nor carry off soluble matter from it, but it is itself filtered of whatever matter it contains that can be appropriated by the roots of the grass.

The width of the levels that may be irrigated is very irregular, and depending greatly upon the character of the surface. The larger the breadth the cheaper the process of preparing the surface, because the expense of forming the embankments, canals, sluices, and drains, is divided over a larger number of acres, and the cost per acre is diminished. It is cheaper to enclose a large area—100 acres for instance—although the works may be heavier and more costly, than a smaller one of 10 acres with much lighter works. In laying out water meadows, this consideration should not be neglected, and the largest area possible should be enclosed. Some of the dikes enclosing the English and Italian water meadows are not less than 20 feet in hight, but hundreds of acres are brought under irrigation by them. In such cases the works are massive, costly, and built to last for ages. Smaller meadows may not require embankments of more than one to three feet in hight, and the earth for these may be procured from the drains which carry off the surplus water, and which are necessarily of ample size. In making the banks it will be found the cheapest plan to dig the drains large enough to supply all the earth needed for the banks ; the extra ground used will be of very little importance compared with the expense of bringing earth from a distance for the construction of the banks.

A water meadow, or at least each section of a meadow in one enclosure, must necessarily be carefully leveled. The most perfect meadow is one that has a perfectly level surface between the banks, so that it can be covered evenly with six inches of water. The water may be flowed over the surface of a meadow of this character, and kept upon it, if desired, by closing the outlet at the foot; or the outlet may be opened only so much as to allow a gentle current to pass over the meadow and maintaining the water at its stated depth. Upon level meadows less water may be used than upon meadows having considerable slope. The more water that can be made to pass over the grass, the better, up to the point of the saturation of the soil. The quantity of water that may be used depends upon the inclination of the surface and the quality of the soil.

Where the surface is perfectly level, and of a clayey character, the minimum quantity of water can be used. When the surface slopes so as to reach the extreme inclination practicable for these meadows, and the soil is gravelly, sandy, and porous, with a porous subsoil, then the maximum quantity of water can be used.

An instance is stated by M. Hervé Mangon, in his work already referred to, of the irrigation of meadows in the valleys of the Vosges, Eastern France, in which water is employed to such an extraordinary extent that the total quantity used in a year would cover the soil to a depth of thirteen hundred feet. In another case the quantity of water used between the end of November and the middle of August following, was equal to a total depth of 27 feet. The whole of this time was divided into eight periods of watering. But the locality in which these extreme cases occurred, is one where the meadows are rarely level, and have generally an extreme inclination; the soil is gravelly, being derived from the schistose rocks of the surrounding hills, and is very porous and loose in texture, and the water of the streams is highly charged with sediment and soluble matter, from the decomposed rocks. At least such is the case in the valley of Waldersbach, a locality much visited by travelers on account of its connection with the history of the renowned Father Oberlin, and where the author has seen the grassy hill sides flowed with sheets of water almost approaching the character of cascades, and the level meadows appearing as lakes.

It is by the use of the most liberal supply of water, when the conditions are favorable, that we can cause to pass over a given surface, the greater quantity of nitrogen, phosphates, and other valuable matters, contained in the water and needed by the soil. Irrigation of meadows is thus seen to be by no means a simple drenching of the soil by stagnant water; but, on the contrary, the bringing into active contact with the soil of the largest possible quantity of water surcharged with fertilizing gases, salts, and organic matter.

When the surface slopes, the arrangements of ditches and drains should be made to suit the slope. If the slope is in only one direction, the water can readily be made to flow down the slope from the head to the foot by a system of gates from the canal which passes along the upper part of the meadow. At the foot the water passes into a drain and escapes into the stream, or it is carried from the drain beneath the dividing bank into the next section, and made to flow over the surface of that, as it has already done over that of the previous section. Where the slope is not more than one foot in 100, a considerable depth of water may be maintained upon the surface, as the flow is greatly retarded by the grass. Where the slope is greater than this, the construction of a water meadow must be abandoned, but a modification of it may be used, and a meadow upon which a current of water may be flowed from head to foot without any series of water furrows, may be made and laid out upon the general plan of the true water meadow. But to flood the surface in case of frost would be impossible or injurious, because of the great depth of water that would be required, and Winter irrigation would be either injurious or full of risk. So long as the slope does not much exceed 1 in 100, the meadow may be laid out as a water meadow, if other circumstances favor it; (such as a location upon a stream where there is sufficient fall, to avoid heavy embanking, which however is a rare occurrence); but when the slope exceeds that ratio another system must be adopted. Besides the systems of water meadows, previously described, there are other methods of irrigating grass lands which will be explained hereafter.

The time of continuance and intervals of irrigation of these meadows is of importance. There is always danger that, by reason of a rise of temperature, vegetation may be unduly stimulated. In such a case the water, only insufficiently charged with oxygen, cannot supply the demands of the plants, and they are destroyed unless the water is withdrawn and air supplied, or the temperature lowered by exposure until the stimulus is removed. An interval of a few days is then to be given before the water is again turned on. An irrigation of 10 or 15 days and an interval of five is the general practice. Whenever practicable, a meadow may be divided into three or four portions in the manner before described. Then, in the first case, by flooding two of the divisions, and at the end of five days drawing off the water from the first and turning it upon the third, and after five days more drying the second and flooding the first, and so on continuing, each division would be ten days under water and five days dry. In the second case, if three are under water in succession and one dry, each will be 15 days irrigated and dry for five days. It is impossible to give directions in each case; the experience of the operator must be his guide, and the beginner must exercise caution, learn to know *when* he is right, and then go ahead. A reference to the principles upon which irrigation depends for its good effects, and the circumstances which would make it injurious, must be carefully made whenever there is doubt in the mind of the operator. The general rule already stated, that it is much more common, and easy, to err upon the side of excess, than on the contrary, may be remembered as a caution and safeguard. Still there is less danger from excess in irrigating grass than any other crop.

It might be well to explain at this point that the arrangement here described for making water meadows is exactly applicable to cranberry plantations which require to be flooded. In many cases the slope of such plantations is too great, and consequently there is either an injurious depth of water flowed upon the vines, or the water is not in sufficient supply to permit the covering of the upper portion of the field, and the expense of making the necessary high banks is too onerous. By laying out the meadow as is shown in profile at fig. 57, and in plan at fig. 58, each plot can be flooded to a moderate and sufficient depth with the expenditure of a minimum quantity of water. The cost of making several low banks and smaller drains is not more than that of making one high bank and wide deep drain, and the crop is not injured by an excessive depth of water.

CHAPTER XIII.

IRRIGATION OF MEADOWS AND PASTURES.

While the irrigation of grass land, situated in a river bottom, and having either a level surface or one with but very little slope, as has been described in the previous chapter, is an easy matter, and when the supply of water is ample is the most effective method of making a water meadow, yet the proportion of farms possessing the requisite facilities for a water meadow is comparatively small. Where therefore, there is but a small supply of water and no broad level space of ground, the meadow must be made in some other manner than that previously described. But in all cases, whatever may be the character of the surface, when there is a supply of water flowing above the level of the ground to be watered, an irrigated meadow may be made. It may be a level piece of land, or a piece sloping in one or two directions, or an irregular surface having meandering slopes, or a hill side so steep that wagons cannot be used upon it, any of these may be brought under irrigation, if there is the requisite supply of water.

In preparing meadows for irrigation, the first consideration is the selection of the ground. In this is included the supply of water. It may be that the area that can be covered by the water is too small to return a fair profit on the venture, or the supply of water may be too small for the area upon which it is to be spread. Close calculations should therefore be made, the supply closely measured and the needs accurately estimated. The first cost of preparing the surface being almost the whole expense to be incurred and this being less in proportion as the area increases, it is a measure of economy to spread the water over as much space as possible. If the water is sufficient to flow one acre in one day, by dividing the land into twelve plots and irrigating one each day in succession, the whole may be brought under the improvement.

Upon level lands or those which have but little slope, and that in only one direction, the preparation of the surface is very easy and simple. In this case the irrigation will be by narrow channels, or ditches sodded or sown over with grass which will offer no obstacle to the mower when the crop is cut. The form of the distributing ditches will be of a very obtuse angle, or a light depression of the surface sufficient to confine the current of water which will flow over its edge or edges, and spread in a thin sheet over the surface; slowly sinking into the ground or finding its way into the drain, either by percolation through the soil or by surface flow at the foot of the field or plot. The form of the furrow is seen at fig. 62. It may be two feet wide and four inches deep. As there is no loss of crop in this case, the space occupied by these furrows is of no consideration; the wider and

Fig. 62.—FORM OF WATER FURROW.

shallower they are, however, the more permanent they will be, and the less subject to injury by trampling, should the meadow ever be pastured. The arrangement of the meadow would then be a main supply-canal, so located that the water may be diverted from it to supply any of the subordinate feeders in turn by means of the distribut-

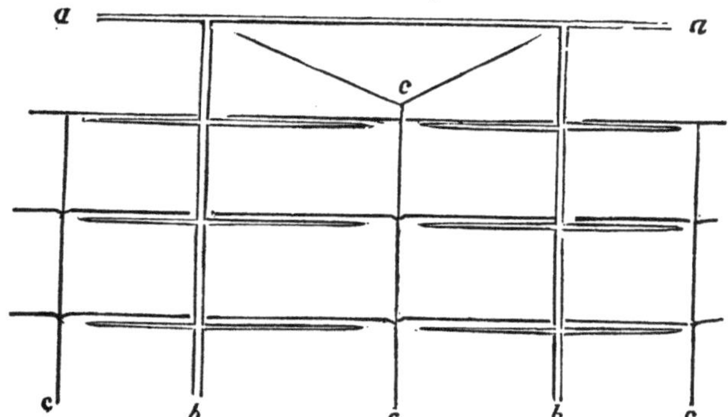

Fig. 63.—PLAN OF IRRIGATED MEADOW.

ing canals. See fig. 63; *a, a,* being the main canal; *b, b,* distributing canals; *c, c,* drains. The flow may be diverted by means of the hand-gates already described, or by placing obstructions in the main canal, such as bricks or sods, as shown in fig. 64.

Where there is an inclination of the surface insufficient to amount to what would be called a slope, a somewhat different arrangement would be required. See fig. 65. The water would be taken from the supply canals and diverted into a feeder to be carried in a diagonal direction across the plot, from which the distributing furrows

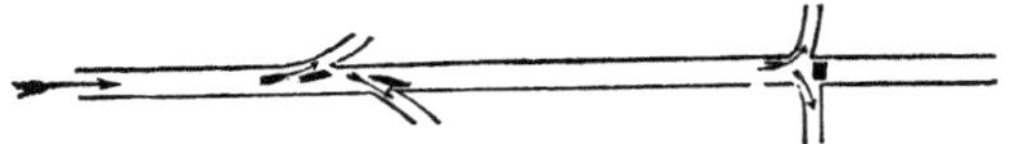

Fig. 64.—DIVERTING THE FLOW.

would be carried. The overflow from the distributing furrows would even spread over the ground down the inclination. The triangular spaces below the junction of the distributing furrows with the feeders, are watered by means of small reflex furrows, which gather some of the overflow from the distributing furrows and carry it back toward the feeder.

This system of irrigated meadows is applicable to numerous and varied circumstances. It may be adopted in cases where the surface is level, or where the inclina-

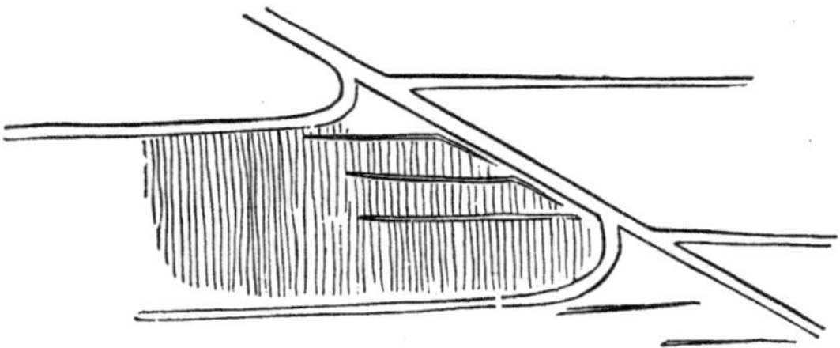

Fig. 65.—FORM OF FURROW FOR AN INCLINED FIELD.

tion is slight but regular, and where the supply of water is not sufficient to permit of flooding or may be in minimum quantity. It may also be adopted in those cases where the surface is a plane of which the slope is moderate in one direction, so that the distributing furrows may be carried on a level, and in a nearly straight direction; or upon nearly all surfaces which will admit the use of a mowing machine. It is particularly adapted to many cases where the land in river bottoms has been injured by the washing of freshets, and a bare surface of sand or gravel has been left, upon which grass cannot now be grown because of the absence of soil. This last case, however, would more properly come under another head, and will be treated in the proper place hereafter.

In preparing the surface of an irrigated meadow, the ground should be plowed without open or back furrows. There may be exceptions to this rule, where the ground is to be laid out in plots for successive irrigation, or where the surface is a dead level. In the former case, the ground may be plowed in broad flat lands, each land forming one plot, of which the open furrow will be the center, and the feeder for the distributing furrows. In the latter case the ground will be plowed in narrower lands, with a rise from side to center of not less than 6 inches to 100 feet; the back furrow or the ridge will be the place for the distributing canal, and the open furrow will be the drain. This will in fact be an extended application of the system of beds heretofore described as applied to gardens. The best implement for this work is the swivel plow, with which the furrows may be all laid the same way over the whole field. The plowing is to be carefully and evenly done, and as deeply as possible. No "balks" must be made, the furrows must be straight, and no trash, weeds, or coarse manure, are to be plowed under, that in rotting would leave depressions of the surface. Two or three plowed crops, or a Summer fallow, might be first taken, so that the surface may be made smooth and level. If there are hollows and knolls, the latter must be leveled and the former filled up. This can be done, in part, with the harrow, and in part with the scraper. The scraper for this purpose may be a plank, at the lower end of which a strip of wide band-iron or saw-plate is fastened. A pair of plow handles are fixed behind, with which it is guided, and a pole or a chain fastened to rings in front, by which it is drawn by a team of oxen or horses. See fig. 66. The common horse-shovel may be used where it is available, and where considerable earth is to be moved, but the plank scraper will make an effective leveler of the ground. The surface is to be rolled and harrowed alternately and repeatedly. Upon the care

Fig. 66.—THE SCRAPER.

and completeness with which this work is done the after value of the meadow will depend. When the surface is prepared, the seed may be sown before the canals and ditches are dug, lest the water should disturb the earth before it is covered with grass and bound together by the roots.

When a surface already level, but without soil sufficient to bear a crop of grass without help, is to be improved by irrigation, the grass seed is sown after flooding, and while the ground is moist, and is left until what will sprout and grow has done so. The water is then turned on to the surface, very gradually, and allowed to flow for 24 hours, when the supply is shut off, and what is upon the surface is permitted to sink into the ground, or flow gradually away. This is repeated, more seed being sown each year, and water being let on whenever it is more than usually charged with solid matter. At every watering some deposit is left, and as the grass increases in growth, more of this solid matter will be arrested, until in a few seasons a sod will be formed, and the meadow begin to yield crops. This method consumes a great quantity of water, but is very usefully applied where there is a stream that is charged with mud or silt after every heavy rain.

When the surface of the plowed meadow is ready for the water, the canals are laid out, with a fall of not more than one foot in 1,000. Whatever is lost in the fall reduces the area that may be watered. The sods are removed carefully from the surface where the canal is dug, and used, after it is completed, to cover the sides. Being cut into pieces, and the pieces placed here and there upon the sides, the intermediate spaces are sown with seed, and the gaps are soon filled. The distributing furrows are made in a similar manner. These may be made with a plow by turning a furrow-slice, in exactly the line laid out, on the opposite side of the furrow from which the water is to overflow. Fig. 67. Great care is to be exercised in laying out the

Fig. 67.—METHOD OF PLOWING THE FURROW.

canals and furrows. A builder's level, fixed to the edge of a plank 12 feet in length, of equal width from end to end, having a cross-bar or foot, a foot long, fastened to each end, will make a useful implement for this purpose. One foot being set on the ground in the line of the ditch, the other is moved from one side to the other in the same direction, until the level is found. A peg is driven there to mark the spot, and the level moved further on. It does not require much ingenuity to do this, and any farmer of ordinary intelligence need not fear that he will go wrong if he will only be careful and cautious as he goes along, and takes the precaution to verify his levels by turning the implement, and going back over the line.

Many rough, stony, or swampy pieces of ground already in grass, may be improved without disturbing the surface, by thoroughly draining the subsoil and laying out canals without reference to any particular line, but merely causing them to follow the level in a direction meandering to suit the surface. Hollows should be filled up with earth taken from adjoining elevations, the sod being first removed and then replaced. Waste pieces of land, at present a refuge and nursery for weeds of many kinds, and a detraction to the farms to which they belong, may thus be changed at small cost into land of the most productive kind.

The irrigation of an irregular surface, such as hill sides, although it may need more careful preparation and adjustment of the levels, is no more difficult than that of a perfect level. In fact, there are advantages in favor of the irregular surface which offset the apparently easier irrigation of a dead level. Drainage is an indispensable adjunct of irrigation, and no land is so frequently drained by nature as a hill side, or what is known as rolling land. Generally the simplest methods of surface drainage will be sufficient for lands of considerable slope. The cost of thorough underdraining is therefore saved in the case of a meadow of this character. The water supply, and the character of the canals suitable for irregular surfaces, differ in no respect from those already described. It is in the method of distributing the water, and laying out the furrows, that especial directions are needed.

There are several methods of irrigating lands of this character, which are applicable to our circumstances. Level furrows may be used by which the water is carried in winding directions around the elevations and depressions of the surface, from feeders which are taken from the main supply canal whenever it may be most convenient. To trace the course of these distributing furrows is very easy, if the common level, already described, is used. The course, as thus laid out, will form a succession of angles, the apex of each of which will be marked by a small peg driven in the ground. To pre-

Fig. 68.—LAYING OUT FURROWS.

vent abrasion of the furrows at these angles, gentle curves are to be made from point to point. These curves will conform exactly to the level of the furrow. Fig. 68 illustrates the method of laying out these curves. If the slope is not so great as to permit washing out of the soil, the feeding canals may be carried straight down it. If the slope is too great for this to be done safely, the feeders will meander in the same manner as the furrows, or

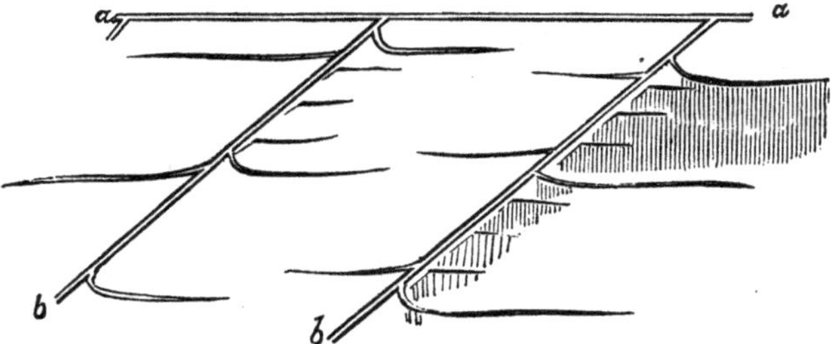

Fig. 69.—FURROWS ON A REGULAR SLOPE.

they may be made to follow a diagonal direction across the slope, so as to bring the fall within proper bounds. The meadow will then appear as in fig. 69, in which *a*, *b*, are the canals or feeders, and the lateral lines the furrows.

But the least troublesome and cheapest method is by inclined furrows carried in the straight lines across the planes of level, and supplied by feeeders carried either directly or diagonally down the slope. The furrows branch both to the right and left from the feeders, and have but very little inclination from the level. They are made to diminish in size from the feeder until each disappears in a point at the extremity. Each feeder with its two lateral ranges of furrows thus appears upon the surface in shape like the backbone of a fish, or what is especially known as "herring-bone shape." Fig. 70 exhibits a plan of a meadow thus laid out. The slope of the field is from top to bottom. The water is received by a main canal,

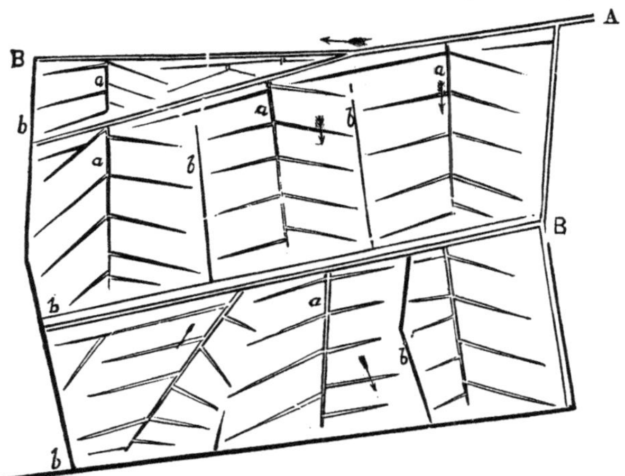

Fig. 70.—FURROWS AND DRAINS FOR IRREGULAR SLOPES.

and is diverted into subsidiary canals, *A*, *B*, and from them into the feeders, *a*, *a*, *a*, and the furrows which branch from them upon each side. The drains are seen at *b*, *b*, *b*. The course of the water is shown by the arrows. The distance between the feeders should be from 100 to 150 feet, which will make the furrows from 50 to 75 feet long, and the latter should be from 15 to 20 feet apart. These distances will be regulated by the character of the soil as to its porosity or retentiveness. The lower extremity of each feeder is closed by a sod or a small gate, and the flow may be regulated or diverted when desirable by the same means at any part of the channel. The drains are placed midway between each feeder, and receive the surplus water, carrying it off at the foot of the meadow. When the water is in flow, notice is to be taken of any portion of the meadow which does not receive a supply, and a special furrow is to be made to remedy the defect.

Drains are not always necessary upon these meadows. If the soil is clay and retentive of moisture, and the slope is slight, they will be indispensable. Where the soil is open and porous, and naturally drained by the subsoil, they may be dispensed with. But attention must be given to so feed the water that it is all used, and not allowed to drown the lower portions of the field. One drain at the foot of the meadow is to be provided in all cases.

Another method of irrigation is adapted to very steep hillsides. This is known as the catch-water system. Hillsides so steep that wagons cannot be taken upon them, may be watered by this system. A stream or canal flowing upon the crest of the hill is dammed, or closed temporarily, by means of a gate. The water then flows over the bank, in a sheet more or less perfect, as the bank has been leveled accurately or otherwise. At some distance down the slope, the water that is not absorbed by the soil is caught in a second canal or ditch, which, when full, overflows and spreads the water upon the section below it. The surplus is caught by a lower canal, and spread as before. This is repeated, until either the water is exhausted or the bottom is reached. If the supply is such that economy is to be exercised, the water may be carried into one of the lower canals by an underground spout of wood, and the meadow be watered in successive portions. The section of a field thus watered is shown in fig. 71. *a*, is the stream, and *b*, *b*, the canals, from which the water flows over the intermediate slopes. The canals in this system follow a perfectly level course, and much care is to be exercised to follow the sinuous course of this level across the meadow. A very safe method is to make the lower side of the canal of plank or slips of board, over the edge of which the water will flow without injury to the canal. The cost of this system of irriga-

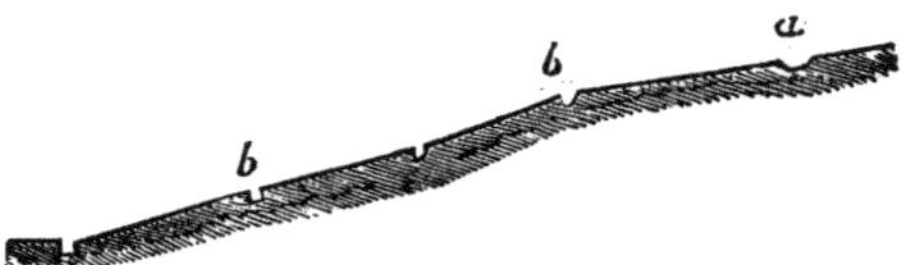

Fig. 71.—CATCH-WATER FURROWS.

tion is frequently not more than $10 per acre. The canals need to be but very small; a furrow that will arrest the flow of water is all that is required, its main office being to restrain the velocity of the water, and to collect it from the numerous streamlets into which it soon gathers, and again spread it in a thin sheet over the whole surface.

Where the surface admits of it, a series of slopes and terraces may be made, which can be irrigated upon this system. See fig. 72. In this case, the slopes may be covered with grass, and the intervals cultivated if desired.

Fig. 72.—SLOPES AND TERRACES.

The water which flows down the slope is caught in the furrow at the foot, and then passes over the terrace on to the next slope. The furrow at the edge of the terrace is needed to retain the water sufficiently to thoroughly irrigate the soil of the terrace, which would possibly otherwise receive less than its share. In this system of irrigation, when the soil is open and porous and the supply of water limited, it will be necessary to puddle the bottom of the canals to prevent loss of water. It may be that the cheapest plan would be to make the bottom and lower side of the canal of boards. In this case, a board of 14 inches in with would form the bottom of the canal, and one of 8 inches the lower side. A canal of this capacity would convey water enough for several acres, and would not be more costly than to puddle or cement the bottom, when clay is not readily at hand.

CHAPTER XIV.

DRAINAGE OF IRRIGATED FIELDS.

The absolute necessity of water to vegetable growth must not be accepted in an unqualified sense. Water is a good and necessary thing, but there may be too much of it, and too much is as fatal to the profitable culture of land as too little. As the circulation of air brings life and vigor to the lungs of an animal, so the circulation of water brings vitality to the roots of a plant. Stagnant water is as fatal to plant growth as stagnant air is to the health and well-being of animals. Therefore irrigation cannot be successfully used without adequate drainage. Sometimes this is naturally provided. Light soils, with gravelly subsoils, may permit the passage of water through them with facility, acting as filters to retain all its fertilizing qualities. Such lands are the most readily adapted to irrigation, and any artificial provision for carrying off the water from them is unnecessary. But there are many lands with retentive surface or subsoil, and others with subsoil practically impermeable to water, that if brought under irrigation must be thoroughly drained, or they will be injured instead of improved, and the character of the vegetation they bear be totally changed. An undrained meadow may be thus, by irrigation, changed into a marsh, and good, though scanty grass be replaced by useless marsh sedges and rushes. Sloping lands may need drainage as much as level lands. Hillsides that have been brought under irrigation, have sometimes discharged their surplus waters at a lower level, where they have gathered and changed a portion of the surface into a quagmire, until drains have been constructed to remedy the evil.

Again, there are cases in which, by a judicious system of drains, a swamp may be reclaimed, and the water, which had been previously a hindrance to cultivation, may be gathered into ditches and used to irrigate a meadow, and yield bounteous crops. Such a case, which actually occurred, may be profitably described. It was a hill-side of fertile clay soil, resting upon a clay slate, from which the soil of a level flat at its foot had been originally derived. Abundant springs broke out upon the hill-side, and after forming marshy spots around them, they disappeared until they again broke out at the foot of the hill, where they gathered and formed a dangerous and impassable swamp. Here were 30 acres of land rendered worthless, and a dangerous trap for any stock that might be tempted to trespass upon its treacherous surface. Hundreds of similar tracts exist where there are hills and valleys.

The reclamation of this tract was a very simple matter. Its outlines are shown at figure 73. A drain was cut near the foot of the hill. See *a*. It was necessary to take this drain to a depth of seven feet before the heaviest springs were cut. At this depth, a flow of water was reached which nearly filled the ditch, and furnished a large stream. The drain was placed, with a view to irrigation of the meadow, a few feet above the level of the flat. It then formed a supply canal from which the flat could be irrigated by means of shallow ditches which led to lateral furrows diverging on each side of the ditches. The surplus water escaped from the foot of the meadow over the bank into a stream, *b*. The plan of the meadow

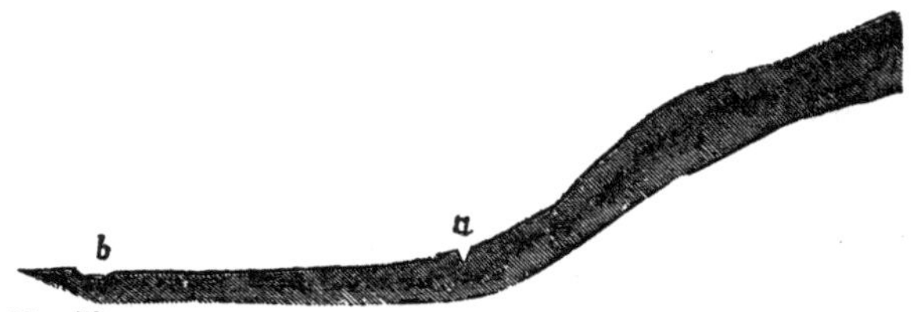

Fig. 73.—SECTION OF A DRAINED HILL AND IRRIGATED FLAT.

is shown at fig. 74. *A* being the hill; *a, a,* the drain, from which the ditches and furrows are led down to the stream, *b, b,* at the foot. By closing the shallow ditches the water could be backed up over the meadow or thrown into lateral ditches. None of these ditches were deep enough to obstruct a mowing machine. It only required

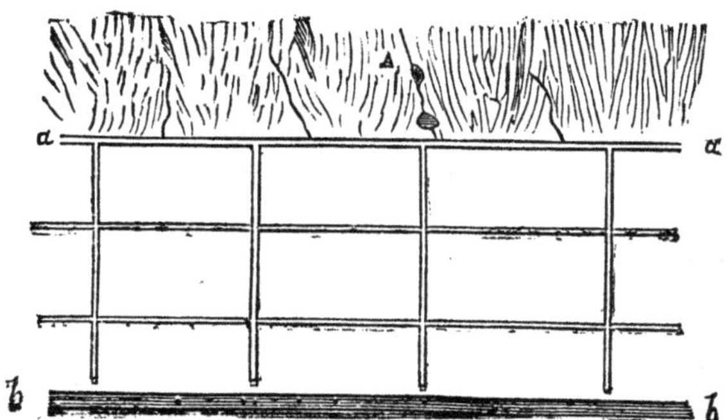

Fig. 74.—PLAN OF THE DRAIN AND FURROWS.

the labor of two men for three months, and the lapse of two years' time, to convert this 30 acres into a dry, arable field of 12 acres, and a meadow of 18 acres, which was covered with grass and clover where, in former years, several cows had been mired and smothered in mud,

It is not the purpose here to treat of drainage with reference to itself alone, but only so far as it may be used in connection with, or as an adjunct to, irrigation. Drainage may be superficial or subterranean. Superficial, or surface drainage, is the simplest. Nothing is needed for its practice but to provide open channels into which the surplus surface water may find its way. As a matter of necessity, these to be perfect must be placed at the lowest levels of the ground to be drained. Besides, they need to be placed in such relation to the distributing furrows of the irrigating system, as to catch the water as soon as it has completely accomplished its purpose, and remove it in the most effective manner. Sufficient description of needed methods has already been given, to make clear the means of doing this. For subterranean, or subsoil drainage, much more elaborate and costly methods are necessary. Not only must expensive ditches be made, and earthen tiles be used, but the arrangement of the drains, with reference to the irrigating ditches or furrows, must be carefully made.

No drain should exist immediately beneath an irrigating ditch, canal or furrow, for the reason that excavated earth cannot be so returned as to be as compact as it laid before. If then a water channel passes across, or along a line of earth, that has been disturbed, a rapid infiltration occurs, the water makes itself a channel, which is rapidly enlarged ; sand or earth is carried into the drains, and the water not only escapes without doing its work, but chokes the drains in a short time. Thus no drain should be made nearer to an irrigating furrow, or canal, than six feet, and no irrigating furrow should terminate at a less distance from the line of a drain, than six feet. The usual arrangement of drains and furrows is shown at figure 75. Here *A, A,* is the main canal ; *A, B, A, C,* the feeders, with the lateral distributing canals or furrows ; *a,* the main drain, which discharges into the outlet, and *c*, *c*, are the small collecting drains. The small drains follow the direction of the greatest slope of the ground.

The system of drains to be adopted will, in all cases, conform to that of the system of canals and furrows. When in perfection the drainage system will be an exact counterpart of that of the irrigation, and so devised as to carry off the water after its service has been performed, and to

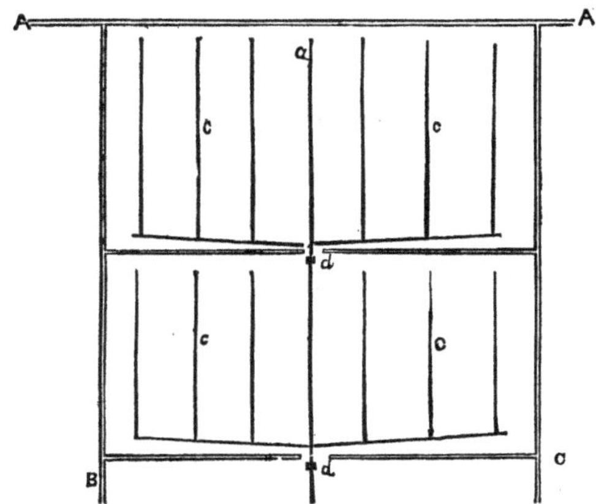

Fig. 75.—MANNER OF SUB-DRAINING AN IRRIGATED MEADOW.

cause it to circulate completely through every portion of the soil occupied by the roots of the grass, after it has been spread completely over the surface. The construction of the drains is in no wise different from that of ordinary tile drains, and therefore needs no description here.

It is sometimes found of great service, consequent upon the frequency with which sand or earthy sediment is carried into the drains, to provide a method of flooding and flushing them. This is called intermittent drainage. It is applied also very advantageously to fields that are subjected to intermittent irrigation, or irrigation by successive portions. It consists in supplying an earthen or wooden pipe, which is set perpendicularly in the ground in the line of the main drain, so that the main drain pipe enters it upon one side, and leaves it upon the other. This pipe thus cuts the main drain at such intervals as may be desirable. It is covered by a cap, and is reached through a covered trap or box, placed on a level with the surface of the ground. This is seen at fig. 76. The proper situations for these pipes are just below the junction of a series of lateral drains, as at *d, d,* in fig. 75. These pipes offer facilities for closing the drain by means of simple contrivances. The most effective of these is a plug or cushion of wood, which fits in the drain leading from the pipe or well. This plug is fastened horizontally to the lower portion of the T-shaped arm. One of the upper cross parts of the T is fixed into a hole or groove in the pipe or well, and a wire is fastened to the other cross part. When the wire is pulled, it moves the lower portion of the T laterally, and draws the plug from the opening of the drain pipe. When the wire is released, the weight of the arm of the T carries the plug to its place again, or the force of the water flowing through the drain carries it and holds it there. This is shown in fig. 76, in which *A*, *B*, is the drain pipe; *a*, the box or trap by which the wire is reached; *b*, the plug with its movable arm, from which a copper wire is carried to the upper box, where it is secured by a ring upon a hook.

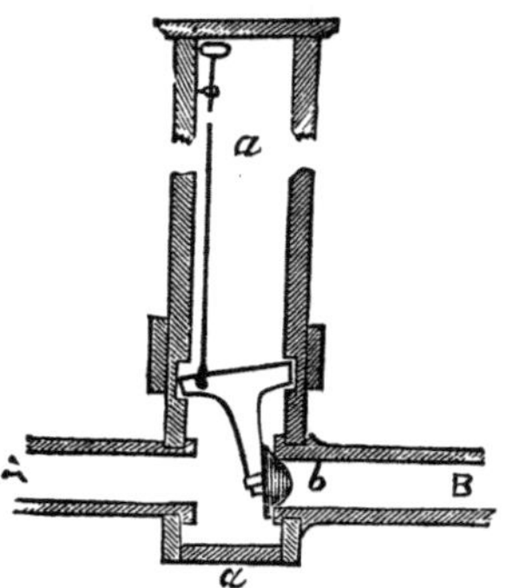

Fig. 76.—PLUG FOR CLOSING DRAIN.

The operation of the contrivance is as follows: When the drain is closed and the flow is stopped, all the drains above the obstacle are charged with water. Water also accumulates in the subsoil and soil, and in fact the whole portion of the field under the influence of the drains, becomes filled with water as completely as may be desired. At any time when the drain may be opened, there is a rush of water through the drains, by which any sediment is effectively carried away, and the drains left free and clear. The operation may be repeated upon each division of the field consecutively from the foot upwards. Instead of the plug above described, an iron rod, having a curved sheet of zinc, or tinned or galvanized iron, attached to the end, see fig. 77, may be used. The curved sheet reaches quite round the well, and when drawn up opens the drain, but when pushed to the bottom, closes it. If the well is square, a slide made to move in grooves may be used to close the drain. The simpler the method, the less risk there is in its use; but the need of permanence of structure is obvious, for if it gets out of order, nothing remains but to take up the well and replace it.

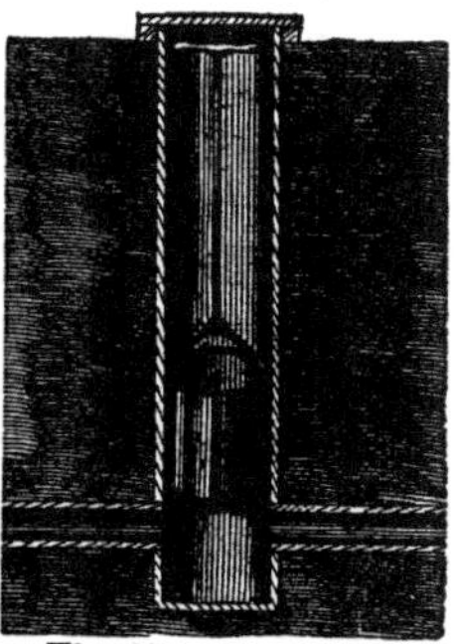

Fig. 77.—A CURVED DRAIN-STOPPER.

Whatever system of drainage is adopted, is immaterial, if the main points here touched upon are provided for. It must not be forgotten, however, that drainage is indispensable, and that except under rare circumstances, thorough subsoil drainage only will be sufficient to meet all the requirements of the case; and that surface drainage may be an unsatisfactory makeshift for the more perfect method. The size of tile used, is one inch for the small drain, two inches for the laterals, and three or four inches, or even larger than that, for the main drains. The size of the main drains should bear a proper proportion to the quantity of water admitted to the field, and it may be that the discharge drain, into which the main drains enter, may need to be six or eight inches in diameter. This question, as to the size of tile, will need careful consideration, because if the size is insufficient, the flow will be retarded, with the inevitable result of sediment and choked drains. As a general rule, the main pipes for irrigated meadows should be twice as large as those used for ordinary drains, as the excess of surplus water at times may be very large. Any system of pipes, that is not equal to the most exacting emergency, will be insufficient, and calculations must be made to meet such an emergency. Before any large expenditure of money or labor is made in laying down drains, which once laid, admit of no remedy except total undoing of the work and relaying the pipes, it would be judicious to consult a capable civil engineer, who could readily make safe calculations as to the size of pipe, the position of the drains, and the number required.

CHAPTER XV

MANAGEMENT OF IRRIGATED FIELDS.

When a field has been successfully irrigated and drained at great expense, it may be seriously injured for want of proper management. To care properly for an irrigated meadow calls for the exercise of tact and skill of no mean character. A few general rules may be laid down for the proper management of irrigated meadows, which will serve to meet the majority of cases, and by modifications of which exceptional cases may be met. The point of chief importance is to avoid pasturing. No hoof should be permitted upon a completely irrigated meadow, unless it be, under certain restrictions, those of sheep. Sheep may be allowed to pasture such a meadow after the last crop of hay has been made, and a sufficient interval has elapsed to thoroughly dry the ground and give the grass a start again. There is no better or cheaper way to fertilize a meadow than this. But if heavy rains occur, the flock should be removed at once, and not admitted until the ground is dry again. Where a tough, thick sod covers the ground, greater latitude may be permitted. There are irrigated meadows in parts of England which possess a sod so dense, and such a heavy growth of grass, that one acre inclosed with hurdles is the regular daily

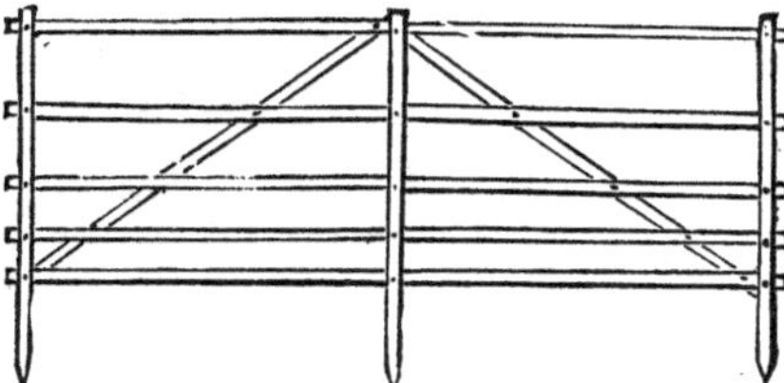

Fig. 78.—FORM OF HURDLE.

allowance of pasture for 1,000 sheep. This is equal to $43\frac{1}{2}$ square feet for each sheep, or a space of 4 by 11 feet only. The droppings of such a flock, so fed, will be a rich and most evenly distributed manuring, and when, as in the cases referred to, the sheep are fed with oil-cake or grain in addition to the pasture, a great increase of fertility results. But it is questionable if we shall ever see such a meadow under our more ardent skies, unless it be by means of irrigation and fertilizing, such as are there in use.

The use of hurdles, for pasturing sheep upon irrigated meadows, is an absolute necessity. Unless confined in this way, sheep will wander over every portion of a field in one day, and picking out some favored spot will remain there, leaving others. The flock should then be confined, in such a space as they may pasture down evenly, and moved daily to a fresh portion. There are various kinds of hurdles used for this purpose. A light hurdle may be made of split poles or laths, three inches in diameter for the uprights, and an inch and a half in diameter for the bars. The upright ends project below for a foot, and are pointed. By driving the pointed ends into the ground with a wooden mallet, the hurdles are kept in place, and

Fig. 79.—LOOSE HURDLE.

standing end to end, form a light portable fence, which can be quickly taken down and set up again. This is seen at fig. 78. Another easily made and portable hurdle consists of a pole or scantling, 10 feet long, bored with holes, alternately in opposite directions, and 12 inches apart. Stakes five feet long are put through these holes, making a hurdle with a cross-section like the letter X. See fig. 79. These hurdles are merely placed upon the ground, resting upon the ends of the stakes, and may be rolled over and over from place to place. Set end to end, they form a fence that is not only impenetrable, but is uninviting in appearance to a sheep given to transgress beyond its legitimate bounds. An arrangement is common among shepherds in England by which hurdles are used with great economy in material and labor of removal. A plot of about a square acre is supposed to be inclosed. This may be done by four lines of hurdles, of 200 feet each. Half an acre may be fed by placing the fourth line across from the middle of the second and third lines of hurdles, thus dividing the plot. The second half acre is fed by moving the fourth line to the ends of the second and third. Adjoining plots are fed in the same manner by moving three lines of hurdles, leaving one to be one of the sides of the new plot. This plan is followed until the whole field has been gone over. Fig. 80 exhibits a diagram which shows each plot numbered successively as fed.

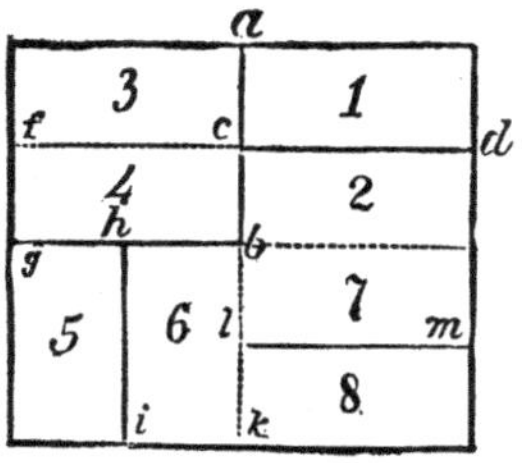

Fig. 80.—PLAN OF SETTING HURDLES.

After a field is pastured, it should be rolled with a smooth, heavy roller. Frequent rolling is very beneficial to an irrigated meadow, smoothing and compacting the surface; but it should be done only when the ground is dry, and in a line with the feeding canals. The roller may be taken across the distributing furrows, when they are properly made, but in no other direction than directly across them. Wagons should be used very carefully upon meadows, and should never be heavily loaded, lest ruts may be cut to the injury of the surface. Wooden shod sleds are preferable to wagons. In case a temporary bridge across a canal or feeder is needed, it may be made by placing a few stout poles from each bank to the bottom of the canal upon the opposite sides, crossing them and placing upon them in a line with the canal a few rails or poles to make a level passage. This leaves a passage for the water in the canal, and by laying two or three poles or rails on the ground at each side of the canal, the edges will be preserved from injury.

Generally, the season of irrigation in our Northern States will be from April to October. As the climate becomes warmer, and as the Southern States are approached, the season will be lengthened at each end, commencing earlier in the Spring and closing later in Autumn. In some localities the season will continue through the Winter. But when the water is warmer than the soil, danger of unseasonably exciting vegetation is to be apprehended where severe late frosts occasionally occur, and must be carefully guarded against, either by suspending the irrigation, or by refraining from drawing off the water from meadows that are entirely submerged, leaving the covering of water as a temporary protection until the danger has passed. The best times for flowing meadows are at night, or on cloudy, calm days when there is little wind, or when it is raining ; windy, clear days or times when the sun is bright, should be avoided. The reason for this is, that the rapid evaporation which would occur on bright sunny or windy days would greatly depress the temperature of the wet soil and retard the growth of the grass. A calm evening is the most favorable time for any irrigation, and nocturnal watering tends to restrain the radiation of heat from the surface, which is active upon calm, clear nights. Intelligent judgment is to be exercised in this regard. When water is applied to a meadow, it is better to give it abundantly rather than sparingly. Generally the temperature of the water is, or should be, higher than that of the subsoil in which the roots exist. A copious irrigation will be sufficient to overcome this lower temperature, and raise it sensibly, with beneficial effect. A more moderate irrigation would, on the contrary, be in danger of producing a contrary effect, because it would be insufficient to overcome the loss consequent upon the increased evaporation.

The nature of the soil needs to be studied in regard to the quantity of water that should be applied, and the periods of its application. A porous soil should be copiously irrigated for short periods and at short intervals. The object is to supply nutriment to the grass, not to cause an excessive filtration through the subsoil, which might carry away valuable fertilizing matter. On a retentive soil the irrigation should be less copious, lest surface exhaustion should occur by washing away valuable soluble matters, and it may be continued for longer periods with longer intervals between them. The usual periods of irrigation are, for 24 to 72 hours, at intervals of four to twelve days. The condition of the soil is the only guide for these.

It is necessary that the irrigator exercise great watchfulness over his field and become acquainted with all its special characteristics, that he may direct the water in this place or that; that he may withdraw here or there; that he may regulate the supply of water according to the condition of the vegetation, or the state of the weather, which may change from day to day. Where irrigation is extensively practiced, a person whose sole attention is directed to the application of the water, ought to be employed. One capable man could attend to several hundred acres, and might earn his season's wages by preventing in time one single mishap, which would probably be overlooked by a person who had the care of many other details upon his mind. Besides, there are frequent occasions, where water is purchased by the inch, in which attention during the night is necessary to prevent waste. This special care is indispensable in the irrigation of field crops, but although not vital to success in case of meadows, is nevertheless of advantage and importance.

Every Autumn the drains, canals, feeders, and furrows should be repaired, sodded, or put in perfect order. The soil is in an unfit condition to be disturbed so early in Spring as might be necessary, and heavy Winter rains might easily devastate a system of imperfect or damaged canals or ditches. Where trees are growing in the meadows, the dead leaves should be carefully raked up and removed, lest the drains be choked with them. Coarse manure should never be spread upon a water meadow. If fertilizers are needed, wood ashes, guano, superphosphate of lime, and plaster only should be applied. In the Spring of the year, after the early floods have passed away, an occasional dressing of one or more of these may be given when thought necessary. If coarse manure must be used upon such a meadow, in the absence of all other fertilizers, it should be spread in the Fall and raked up carefully in the Spring with the horse-rake, leaving no litter upon the field.

Rolling the surface in the Spring, after the ground has become dry, will be imperative. Any inequalities of the surface not thus removed, should be remedied with the shovel, first removing the sod and then replacing it and beating it down firmly. A lawn mower would serve excellently to remove the grass from the distributing furrows, passing up on one side and down on the other. This should be done as frequently as may be needed to keep the current free from obstruction. Otherwise this work should be done with the scythe, but the small cost of a lawn-mower will be amply returned in one season by the saving of time in attending to one acre of meadow. Irrigation should be suspended at least eight days before the crop is mown. The length of this interval will depend somewhat upon circumstances, which may hasten or retard the drying of the ground. If mown by hand, the field need not be so dry as if mown by horse and machine. If the weather is very dry, an irrigation of an hour or two during the previous evening will moisten the grass and greatly facilitate the cutting. Valuable herbage may be encouraged and useless weeds repressed, to a great extent, by the use of superphosphate of lime as an occasional dressing. Excessive watering encourages coarse grasses and sedges, and the growth of these injurious weeds must be carefully guarded against, by care in applying the water and by drainage. The early maturity of the grass necessitates early cutting. The proper time for cutting is before the seed is ripe, and immediately after the blossoming is past. Some of the grasses thrive best when cut before blossoming, and recover the check without loss of time. For a perennial meadow, this is a matter for observation and experience, and is important to study. Re-seeding in part will be occasionally needed. No grass endures indefinitely, and as the herbage dies out, it must be reinforced by new seed. This is to be done by spreading from time to time, when found necessary, a sufficient quantity of that kind of seed which is found to grow most thriftily in the locality, and upon each particular soil. These conditions are so diverse that it would be useless to give even general directions, or attempt to meet them. Each owner of a meadow must in this be a law unto and a judge for himself.

Some species of grass will bear much more cutting than others. One of the best of our common grasses for irrigated meadows is Kentucky Blue-grass. This has been found to submit to frequent watering, and has made a profuse growth, more especially late in the season. A watered meadow, covered almost wholly with this grass, has been in fine condition for cutting in September, and has yielded a good crop of hay as late as the first week in October. After that date the field furnished abundant pasturage until the severe frosts made the herbage unwholesome. This instance occurred in Eastern Pennsylvania. Where meadows are only partially irrigated; they may be pastured freely, so long as the soil is dry and is not "poached" or cut up by the hoofs of cattle or horses. Such meadows should be laid out with broad, shallow water-furrows, so that there will be no danger of the edges being broken down by the trampling of the stock. The late pasturing of meadows by sheep should never be permitted, unless the growth is thick and heavy, as these animals nip the grass very closely and would expose the roots to the frost, endangering the unequal heaving of the surface during the Winter. But as cattle are accustomed to bite here and there, and leave scattered bunches of the rejected herbage, it should be made a business, not to be neglected, to go over these with a mower and level them before the season is closed. The droppings of the cattle should be broken up finely and scattered over the surface before they become frozen. Early in Spring, before any water is given, the meadows should be put into the best condition, the surface cleared of rubbish, and rolled, the ditches and furrows examined and repaired, and the drains cleared if this is needed. It is at this time that any seed or fertilizers that may be thought necessary, should be given. Lastly, the fences should be made perfectly safe. The trespassing of heavy stock upon a newly watered meadow might do very serious and very extensive injury.

When the grass is nearly ready for cutting, no water should be given. In general, the watering should be so timed, that the growth of grass is pushed forward as quickly as possible in its earlier stages, and when the herbage is short. When the ground is well covered and shaded, a good soaking will supply the soil with sufficient moisture to mature the crop of grass. Then ten days or two weeks may elapse between the last watering and the cutting. As soon as the hay is made and removed, water should be given immediately, but never in the day-time, unless the weather is cloudy, or it is raining. In the evening, after sundown, the water may be given, and soon after sunrise it should be turned off, unless it is in very moderate quantity.

When a water meadow is flooded, it is necessary to watch the water, and the moment a white scum is observed to float upon the surface, the water should be drawn off. Meadows that are flooded by streams, should not be watered in time of freshets, if there is any sediment in the water, unless very early in the Spring, or immediately after hay has been made. If the grass is of any considerable length, the suspended matter will be retained amongst it and make it gritty or sandy, and seriously interfere with the cutting. The flooding of a water meadow is preferably done during the Winter, when the solid matter, deposited by the water, is of the greatest value as a fertilizer; the Summer-watering is to supply the needed moisture only, and not to fertilize the crop in the sense of adding manurial matter. Summer irrigation is therefore only moderate in quantity, and an excess of water will be injurious at this season.

The most suitable fertilizers for irrigated meadows are nitrate of soda and Peruvian guano, used alternately, and not mixed together. Where the growth of grass is forced so much as under irrigation, active and soluble fertilizers given in small quantities, and frequently, are required. The proper periods for their application are early in Spring, and immediately after the cutting of the grass; 80 pounds per acre, or half a pound to the square-rod of either, will be a sufficient quantity to apply at once, and the repetition of this top-dressing may be given only when the condition of the grass seems to call for it. Every fifth or sixth year a dressing of lime may be given in Winter, and should be spread upon the snow if possible, (for the preservation of the surface) rather than in any other manner. Where it is known that lime is effective upon the soil, it may be used in the same manner as upon other lands; if used experimentally, 40 bushels per acre may be taken as the normal quantity. The needs of the soil as to fertilizing may be calculated as proportionate to the drafts made upon it. Where hay is removed and the meadow is not pastured, at least the amount of fertilizing matter mentioned above will be required, or even more, if the grass crops are heavy and of good quality. If sheep are pastured upon a meadow in the daytime, and fed at night in yards, with bran, grain, or any extra food, or if dairy cows are so pastured and fed, the need for fertilizers will be small, or none may be required. A reasonable consideration should be given to this point, which will be an easy matter for the intelligent farmer.

In pasturing these meadows, it will be best to stock them closely, and use only a portion at a time, that the grass may be eaten off clean, and not trampled down. By dividing the meadow into sections, it will be easy to arrange for pasturing one part, while the others are either under water or in different stages of growth. As a general rule it is advisable not to pasture sheep freely upon watered meadows, unless they are fed for the butcher. When fed for fattening, they make very rapid growth; the lush herbage causes an excessive secretion of bile, which at first assists greatly in the formation of a high-colored fat, but after this favorable stage is passed, the blood may, and will probably, become affected and inflammatory disease appear; or the sheep will almost certainly become infected with flukes, and the rot will inevitably result. Only experienced sheep owners should make use of pastures irrigated by streams, and they should be watchful not to overpass the point of safe exposure to the dangerous feeding. When the meadows are watered from wells, through pipes, as described on page 55, this caution may not be applicable.

CHAPTER XVI.

IRRIGATION OF ARABLE LANDS.

Few of us ever consider that the larger portion of the arable surface of the United States is doomed to comparative sterility, unless brought under systematic and permanent irrigation. West of the 100th meridian of longitude, almost to the shores of the Pacific Ocean, and from our southern to our northern boundary, stretches a vast tract of land, rich in every element of fertility but moisture, and useless for the purposes of agriculture in its present condition. But while the immense tract is arid in its climate, and for all practical purposes it may be said to be absolutely rainless, yet there flows, across or beneath its surface, the water-shed of a vast and intricate range of mountains, snow-clad during a part or the whole of the year, and which divides it into two portions. It needs but to capture this water, and spread it over the surface, to insure abundant and certain harvests. It may surprise a farmer, used to depend upon the changeful seasons of the Eastern part of the country, to learn that upon these arid lands there may be grown luxurious crops of grass, grain or roots, with the greatest certainty ; that in this climate, the farmer who has brought the waters beneath his yoke, has secured literally and naturally the fulfillment of the promise, that seed-time and harvest should never more fail, while he himself enjoys it only in part and accidentally, and occasionally fails completely to realize it. But this is the fact, for drouth and aridity are entirely subjugated by means of irrigation, and are, strangely enough, only sources of anxiety and loss in those districts were rain falls, and the farmer is subject to conditions of climate which he can neither foresee nor control. Seed-time and harvest are only sure where irrigation is systematically used by the cultivator.

But the irrigation of lands of the character under consideration, can only be profitably undertaken by the combined effort of a community. The necessary engineering works, such as dams, canals, sluices, water-ways, and aqueducts, can only be constructed by means of ample capital, and for the use of numerous farmers, cultivating in the aggregate many thousands of acres. In such cases, the total cost divided among the farms to be irrigated, would leave for each one a sum far less than that needed to clear a farm of equal size from the forest. The actual cost of irrigating works of a permanent character, has been found to range from so small a sum as $1 per acre, upward. That is, a community of farmers, numbering some hundreds, may construct the necessary dams, canals, sluices and feed-gates to irrigate 10,000 to 50,000 acres of land, at a total cost not to exceed $5 per acre, where the conditions of water supply, character of soil, and surface of the land are favorable. To clear an acre of average timber land, will cost $12 to $25 per acre, and the money value of the damage incurred annually, by reason of the stumps and roots which interfere with cultivation, until they have rotted away or have been removed with infinite labor, may easily amount to $20 per acre more. To irrigate a farm permanently, may then cost but one-eighth of the sum necessary to clear it of timber. This estimate will allow of substantially constructed works, which will require but little repair, or renewal, to keep them in permanently good condition. Large tracts of land have been supplied with water for irrigation, at a much less cost than this, in some cases even so low as 25 to 50 cents per acre; but this cost covers only the construction of the main supply ditch, and not the interior ditches, which, to be permanent, should be well laid out, and properly constructed. It has been sufficiently well shown, however, that a supply of water for irrigation can be brought to and spread over a farm upon our dry plains, at a total expenditure of capital per acre not any greater than the annual rent paid per acre for irrigating water in European countries. It is true that we have cheap land upon which to construct the ditches, and that so far, for want of preoccupation of the land, the course of the canal has been made to follow the meanderings of the line of grade chosen, and to save the cost of expensive aqueducts across valleys and depressions, and that economy in the use of the water has not been an object of serious consideration; but it may be some years or even centuries before the cost of irrigation with us can reach $5 to $10 per acre yearly, which is the cost of water supplied by canals of Europe.

This cheapness of irrigation must undoubtedly give a great impetus to the settlement of lands upon the plains and great valleys, so soon as those who are now occasionally brought to the verge of ruin, by the failure of their crops, can be brought to understand its cheapness, its ease of application, and the certainty of crops secured by its means. The numerous settlements that have been made in California, Colorado, Utah, and other localities, and the success which has attended these pioneer efforts, in spite of all the drawbacks incident to a want of knowledge of the peculiarities of the climate and the soil, and inexperience in the art of irrigation, will tend greatly to attract men toward new enterprises in this direction. The evident advantages of a system of agriculture, in which water can be supplied to the fields at will, and dependence upon a fickle and uncertain or arid climate is avoided, will have numerous attractions to men who have seen the fruits of their labor perish, year after year, either by drouth or excess of rain; and as soon as trustworthy and exact information can be procured, thousands of settlers will avail themselves of the benefits offered to them by a cheap, fertile soil, clear skies, genial climate, and water constantly at hand, and under the most perfect control.

Irrigation of land is an art that has existed for many centuries previous to any authentic written history. The traditions of the Chinese people are very ancient, and irrigation is mentioned in their most ancient traditional history, as being extensively practiced. In Egypt, Syria, and the ancient kingdoms of Eastern Asia, agriculture depended almost wholly upon irrigation, and still so depends in these lands where the people have survived the political changes of thousands of years. Virgil in his rural poems thus describes exactly the processes which are followed now. "He leads the stream and flowing rivulets, to the growing corn, and when the burnt field dries up, the herbs dying, he leads the water and cools the parched fields with rills." The irrigation of gardens, vineyards, and fields, is frequently referred to in the Scriptures, one of the earliest books speaks of it and one of the prophets refers to "furrows of the plantation." And so agriculture has continued to the present day, the necessities of the majority of the cultivators of the soil in the Eastern hemisphere, and the natural opportunities possessed by them, combining to render the system vital to their existense. When the Spaniards occupied the new found continent, they introduced their system of irrigation wherever the dryness of the climate demanded it. In Chili, Peru, Central America, and Mexico, the canals and ditches made by the early Spanish settlers remain, and many are still in use; the systems adopted in California, Texas, New Mexico, and Colorado, are mainly copied from the ancient models. It is hardly necessary to say that these models are not of the best construction, nor at all satisfactory to the engineer of the present day, but they are of cheap and easy construction.

The settlement of the drier regions of our territory, adds another instance to those of past history, of the reclamation of deserts by irrigation. It will be of interest to glance over what has already been d

ways in this

way, before considering the possibilities of the future. The actual history of irrigation in the United States begins with the occupation of Utah by the Mormons in 1846. At that time the territory was a waste of barren land and sage brush. In 1868, twenty two years after the first settlement of Salt Lake valley, 93,799 acres of land were under irrigation at an expense of nearly $250,000, and works were in course of construction which, when completed, would greatly enlarge the area of land under cultivation. With the exception of the continuance of some of the irrigation works constructed by the Spaniards in Texas, New Mexico, and California, a hundred and fifty years ago, and which have been in use up to the time when the territory came into the possession of the United States, but little was done in the way of irrigation, until the occupation of Colorado and the adjacent territories, when these were rendered accessible by the opening of the Pacific railroads. In the course of a few years a great impetus was given to the settlement of lands adjacent to the rivers, and which could be brought under irrigation, and several extensive works were constructed. Amongst these may be mentioned the Platte River canal, 24 miles long, irrigating 50,000 acres of land, and supplying the city of Denver. Originally, the canal was 10 feet wide and 2 feet deep at the head, but has been enlarged to 18 feet in width and 3 feet in depth. The fall is irregular, varying from 6 feet to 18 inches per mile. The cost was $100,000; a very excessive amount, but probably unavoidably so on account of its unscientific and wasteful mode of construction.

The Table Mountain Ditch Company Canal, near Golden City, is nearly 20 miles long; 12 to 15 feet wide at the surface, and 6 feet at the bottom, and 2 feet deep. The fall is 19 feet to the mile, in portions, and in consequence of this excessive slope the ditch is destroying itself very rapidly. A branch is 2½ miles long. The cost was, (in 1865), about $15,000. From its faulty contruction, it was dear, and will be costly to maintain. The charge for water is $1.50 per inch, per year; equal to about $1 per acre, yearly.

The Farmers' Ditch, also near Golden City, is 11 miles long, 8 to 12 feet wide on the surface, and 6 feet at the bottom, and 18 inches deep. Its cost was $10,000; it supplies nearly 40,000 acres, at a cost of about $1 per acre, yearly.

At Greeley, on the Cache la Poudre, there are two irrigating canals, one on the south side of the Cache la Poudre river, 10 miles long, which supplies the town and adjacent farming land, it is 15 feet wide on the bottom for 8 miles, has a fall of 3 feet to the mile, and the water is usually run 3 feet deep in the irrigating season. This canal has cost about $15,000. The main canal is 32 miles long, and is taken out of the Cache la Poudre river, 15 miles west of Greeley, on the north side of the river. It waters over 20,000 acres of land, of which 10,000 have been brought under cultivation. It is 25 feet wide on the bottom for 3 miles, the next 5 miles it is 24 feet wide, 20 feet wide at the end of the 20th mile, and gradually decreases to 10 feet at the 30th mile. It is 4 feet deep to the 20th mile; its fall is 3 feet to the mile, velocity 3 miles per hour, or 4 and $^{40}|_{100}$ feet per second, slope of banks 1 to 1; total cost, including dam in river, $60,000. The sectional area of the portion that is 24 feet wide on the bottom is 112 feet, or 16,128 square inches, and having a velocity of 3 miles per hour, and it being generally considered that one inch of water is sufficient for each acre under cultivation, this canal is large enough to water 16,000 acres. Each owner of an 80 acre lot under this canal has now paid $250 for his water right, which belongs to the land as a perpetual easement, and smaller and larger lots have paid in proportion. The canal is kept in repair, and a man paid for superintending it during the irrigating season, by a tax on each 80 acres of $8 to $12 annually. The superintendent measures the water into each "lateral" ditch along the line of the main canal, according to the number of water rights paid for in any year, and the farmers divide it from the small ditches themselves, according to what each is entitled to. Usually the farmers taking water from one "lateral" form a company and build their main and sub-laterals, and deliver to each his just proportion of water. Some of the laterals are 4 miles long and have cost over $1,000. The whole system is working satisfactorily to all, and the land is constantly appreciating in value, as the amount of land that will eventually be brought under cultivation is limited to the amount of water in the streams. Probably not more than two million of the 67 million acres in this State can possibly be farmed, as the combined sectional area of all the streams at "high water" is not over 1,500,000 inches, with a velocity of 3 miles per hour, and on an average it takes one inch of water running at that rate for each acre under cultivation. For instance, a farmer having 100 acres in cultivation, gets 100 inches of water with this velocity, and he can get over, or water his crop in about 10 or 12 days. Usually wheat is watered here but two or three times, as there is rain or snow enough in the Spring, (April or May), to bring it up so that it will cover the ground. Corn, potatoes, and other late crops are watered oftener, but require less per acre than wheat. The above particulars are given by Mr. J. D. Buckley, engineer of the Greeley Colony.

The Canal of The Saint Louis Western Colony, at Evans, with its branches, is 40 miles long, 10 feet wide at the bottom, with slopes of 1½ to 1, (or 18 inches horizontal to 12 feet perpendicular), with a water section of 53 square feet, and a fall of 7 feet per mile. The cost of the whole system is less than $25,000 for a total length of 40 miles, and 115,200 acres are covered by it.

A private ditch, belonging to Mr. G. H. Church, of Boulder Co., is 10 miles long, 5 feet wide and 1 foot deep. The fall is excessive, viz.: 13 feet to the mile. It cost $1,000. It is connected with a reservoir, as its supply is not continuous, and a reserve is thus maintained. Forty acres of land, with the farm stock are watered, and a fish pond is supplied by it. The cost of watering is from 50 cents to $1 per acre, according to the character of the season.

The Upper Platte and Bear Creek Ditch, is owned by a company in Arapahoe Co. It is 5 miles long, 16 feet wide, and 20 inches deep at the head, diminishing towards the foot. The cost of maintenance, which is assessed yearly upon the owners, averages $30 to $35 for 144 square inches of water, or a supply sufficient for 150 or 160 acres. Interest on the original cost must be added to this annual charge, to reach the yearly cost of watering. No information as to the original cost has been given. There are many other irrigation works, constructed either by joint effort or by incorporated companies, who lease the water at a remunerative yearly rent. These rents vary from $1.50 to $3.00 an acre per year for each square inch, which is equal to $1 to $2 per acre of land watered. The cost of the manipulation of the water, after it is received by the farmer, will obviously vary with the character of the crops. On the average 50 cents per acre, annually, will cover all expenses of distribution.

As an instance of what has been and may be done in localities where partial irrigation may be usefully applied, a case which occurred in the Arkansas valley, in Central Kansas, may be cited. Here is a broad, level, fertile valley, some miles in width, with gently rising table lands on either flank. Flowing through the center is the Arkansas river, a broad, magnificent stream, which neither floods nor dwindles in volume in the whole year. For several hundred miles after it issues from the mountains it flows through rich, level bottoms, in Colorado and Western Kansas, most of which are too dry for cultivation without irrigation, and now afford only pasturage. In Central Kansas it passes through a rich and beautiful country, now well populated, on the verge of the dry country, but within the arable region. At Hutchinson, in Reno Co., the enterprising inhabitants have cut a canal from the river, for a length of two miles, for the purpose of providing water power for factories, and mills. The fall of the river is 8 feet per mile, which is sufficient to carry the water in the course of a few miles on to the high uplands, and to water these as well as the broad valley. At present there is no intention of using the water for irrigation, but should it become necessary or desirable, it is here shown that an inexhaustible supply of water can be obtained at nominal expense to supply every need of the farmer in the dryest seasons. Also it is clear that the whole of this grand valley may be made available for farms. This is one instance only of what may, and in time undoubtedly will, be done in many places where there is only a partial and occasional use for water.

Irrigation in California has, so far, been done by individual enterprise. In 1871, there were 915 irrigating ditches, supplying only 90,000 acres of land, or on an average, but 100 acres to each ditch. The ditches, with few exceptions, are rude affairs, and of inconsiderable length. The exceptions are as follows : The San Joaquin and Kings River Canal Company, is 38½ miles long, and is supplied by the San Joaquin river. It is 55 feet wide, four feet deep, with a fall of one foot to the mile. 15,000 acres are irrigated by this, and cultivated in wheat, barley and alfalfa, and water enough for 60,000 acres more can be supplied. The extension of the canal 40 miles further, is proposed, by which 325,000 (?) acres can be irrigated. The cost so far is stated to be $500,000, (an enormously excessive cost under any circumstances), and the income, in 1873, for water rent, was less than $10,000. It is evident that for some reason, probably inexperience, and poor engineering, the cost of this canal has been ruinously great. The Kings River Company Canal when completed, is expected to water 300,000 acres (?). There seems to be a serious error in the stated capacity of these canals as will be explained in a future chapter. It is 30 feet wide, 3 feet deep, with a fall of one foot to the mile. In the same valley the canal of Messrs. Chapman, Miller and Lux, taps the San Joaquin river, and runs 30 miles down the valley, supplying 30,000 acres. It is 35 feet wide, 3 feet deep, and falls one foot to the mile. Another canal, owned by Friedlander & Co., takes water from the Fresno river, at the foot hills of the same valley, and supplies 40,000 acres. This ditch is 10 miles long, 40 feet wide, and has a fall of about 10 inches to the mile. A reservoir, connected with the canal is a mile and a half long, 100 feet wide and 6 feet deep. Numerous farms, gardens, and orchards are irrigated by the smaller ditches, and some by wells. San Francisco is chiefly supplied with vegetables from irrigated gardens, many of which are cultivated by Chinese. A small-fruit plantation of 8 acres is watered by a 4½ horse-power engine, from a well. In all the instances referred to, irrigation is successful and profitable. But in California, while irrigation is as yet in embryo, its possibilities are immense. The interests involved, however, are so vast and complicated, the mining interests clashing seriously with those of the farmers, that legislation will undoubtedly need to be invoked before such measures, as will be satisfactory and effective, can be applied to the gigantic natural facilities and opportunities afforded in the valleys of this State.

This naturally leads to the consideration of the ownership of the water, for from this question will probably arise much difficulty and litigation. It is a new element, depending at present upon the principles of common law; no statutory provisions having as yet been made to meet the necessarily involved interests which will be affected by it. Perhaps the only decision which relates to this is cited in the Massachusetts Agricultural Report of 1872, as follows:

"It has sometimes been made a question whether a riparian proprietor can direct water from a running stream for purposes of irrigation.

"The language of the Court as best defining the principles governing this subject is as follows, to wit: That an individual owning a spring on his land, from which water flows in a current through his neighbor's land, would have the right to use the whole of it, if necessary, to satisfy his natural wants. He may consume all the water for his domestic purposes, including water for his stock. If he desires to use it for irrigation, and there is a lower proprietor to whom its use is essential to supply his natural wants, or for his stock, he must use the water so as to leave enough for such lower proprietor. Where the stream is small and does not supply more than sufficient to answer the natural wants of the different proprietors living on it, none of the proprietors can use the water for irrigation or manufactures."

This is so clearly inadequate to meet the urgent necessities of the case, that the immediate attention of Congress, and the various State Legislatures, is peremptorily called for. Fortunately, a beginning has been made, and a Commission was organized by an Act of Congress, approved March 3, 1873, to examine the great valleys of California, with reference to the construction of a system of irrigation. The report of this Commission is published in the yearly volume of the Department of Agriculture for 1874. The conclusions reached may be seriously questioned in many points, but on the whole are, as might have been expected, favorable both to the profitableness and feasibility of irrigation works, and to the interference of the National and State governments, and their control over the distribution of the water. In favor of government control there is both reason and precedent. By no other authority could the conflicting interests of miners, agriculturists, and owners of land to be injured or benefited by the enterprise, be properly reconciled. In Europe, the supreme control is exercised by, and the ownership of the water vested in, the State. The French government in 1669, by special law, reserved the ownership of all rivers and streams, and grants concessions to irrigating companies under restrictions. In Italy, the State has always exercised this ownership, and in Venice the springs, and even the rainfall, so far as it can be stored in reservoirs, have been held to be public property. In India, the springs and rainfall are accumulated in reservoirs, controlled by the government, and the river systems are also owned by it; not only this, but the details of the distribution of the water are also directed by government officials. This is made necessary, however, by the utter incapacity of the ignorant inhabitants to manage anything for themselves, that calls for more than a very low degree of intelligence. Lest, however, it might be urged that government ownership and supervision, is likely to lead to failure, the actual results attained in India may be very properly here cited. During recent years, the British Government has spent about $70,000,000 in irrigating works, and others are in progress of construction which will require half as much more to complete them. In almost every instance the investments have been profitable, and in some cases enormously so, both in the way of water rent, and in service to the cultivators of the soil. The total annual revenue to the government from the works, is more than $5,000,000, or $7\frac{3}{4}$ per cent on the cost. In one case only has there been a loss. The capital expended in the largest works, and the annual revenue from them, is given in the following table, which is derived from the official reports of the East Indian Government:

	Capital invested.	*Annual revenue.*
North Western Provinces	$17,827,225	5 1/4 per cent.
Punjaub	15,671,000	5 "
Madras	9,467,200	22 3/4 "
Bombay and Sind	11,113,940	12 "
Ganges Canal	14,400,890	4 1/2 "
Eastern Jumna Canal	2,350,000	11 1/4 "
Western " "	6,532,000	7 1/2 "
Godavery Delta Works	3,418,525	39 3/4 "
Kistnah " "	2,337,135	13 1/4 "
Canvery " "	1,468,000	36 1/2 "
Sind Inundation Canal	5,930,000	18 1/2 "

The revenue to the government is the least portion of the profit derived from these works. The profit to the people themselves, amounts to a vastly greater sum, one, in fact, the amount of which is not to be computed in money; for the famine, of frequent occurrence before the completion of these works, destroyed thousands of human lives, and caused thousands of square miles of fertile land to be abandoned to grow up to jungle. In 1860, the Ganges canal preserved grain crops from destruction, which fed a million of people; in 1874 the Soave canal saved the crops over a large territory, which would otherwise have been devastated by drouth, and many of the newer works, water regions which have heretofore been visited with some of the most destructive famines mentioned in history. And the whole of this work has been undertaken and successfully managed by the government.

Economy in the use of the water, and in the construction of the works also, calls for such extended surveys, perhaps over hundreds of miles of territory, that no private persons, nor associated companies, could possibly perform them, unless they were endowed with legalized monopolies or exclusive rights; and in the light of past experience with huge chartered corporations, farmers could not wisely submit to have their interests—so vital in this case—placed in such keeping. The experience already gathered in the case of the Cavour canal in Italy, proves that a chartered company is a most unsafe trustee for the interests of the persons most nearly concerned in an irrigating canal. In that case, while fortunes were made by speculators, the work was a failure, and the government was forced to interfere and purchase the canal in the end.

It is, however, out of place to argue this question here, and it is left for the consideration of those interested, who will readily perceive the necessity for the course here indicated.

The cost of irrigation has been very clearly shown by the successful enterprises in Colorado. It must be remembered, however, that nothing is paid for the water itself, and only the expense of bringing it to the fields is included in the figures given. As a rule, the more extensive the works, and the greater area brought under irrigation, the less is the cost per acre. But it is hardly probable that, in any case, the annual cost per acre, can be brought below an average of $1 to $2 per acre. In many cases the cost may be more than this, but even then, the profit to the owner of the land will be many times greater than the cost incurred. Lands that have gone a begging at $5 per acre, in parts of California, and have indeed been practically useless while without water, have been purchased eagerly at $25 to $50 per acre, as soon as a supply of water has been brought to them. In general, the extra value added to land by irrigation, varies from $25 per acre up to several hundred dollars. In some portions of Europe, land is by irrigation increased in salable value, five to ten-fold.

The charge for water in France, varies from $6 to $7 per acre, annually; one cubic foot per second, being used for 70 acres, and $450 being paid per cubic foot per second, during the season. No water is permitted to be given away, although the purchaser may have a surplus. One cubic foot per second, is equal to 72 square inches of water flowing at the rate of 4 miles per hour, or as fast as an active man can walk with ease.

In Spain the Iberian Irrigation Company makes a charge of $7 per acre, for 12 waterings per year, equivalent to a total depth of 33 inches of water over the entire surface irrigated. The canal of this company is a splendid engineering work, it being 28 miles long and costing $600,000. It has a surplus of water, equal to a power of 3,000 horses, which is rented out at $50 per horse power per annum.

In Italy the cost of water varies considerably. In Lombardy, about 1,600,000 acres are irrigated, at an investment of about $20 per acre, or a total of $30,000,000, which is equivalent to $1,250 per cubic foot per second. The increased rental value of the irrigated land is $4,500-000 per annum ; or 15 per cent on the cost of the works. The average cost of the water to the farmer is from $750 to $850 per cubic foot per second, equivalent to $2.50 per acre for maize, $7.50 an acre for meadows, and $20 an acre for rice. A very good idea of the maximum cost of irrigation can be gathered from these figures. The water from some of the canals is purchased by local associations, of farmers or speculators, who distribute it to the irrigators. One of these local associations purchases water from two canals, paying $87 for the cubic foot per second for 714 feet from the first, and selling it out at $96 per foot ; and $65 per foot for 674 feet from the second, and charging $77 per foot. The higher price is charged and paid on account of the valuable fertilizing matter brought down by the water.

These figures, it should be remembered, are fixed by circumstances entirely different from any that are likely to occur in this country. The value of land is higher than with us ; the cost of the canals, aqueducts, bridges, and distributing apparatus, is much higher than would be necessary here, being made with scrupulous care for economy in both water and land ; and the cost of supervision is much higher than would be likely to occur here. Unless costly dams, expensive bridges, and aqueducts, built with a view to the utmost permanency, should be required, there would probably never be any approach made here, to the high cost of water that is experienced in European countries. Some of the European works, now in operation, have been in existence for more than 1,500 years. Others abandoned, but still in serviceable condition, are over 2,000 years old.

The quantity of water needed for irrigation, as has been already explained, varies greatly, and in making estimates of the amount required, for any stated territory, the engineer or irrigator must necessarily study both soil and climate. Where exhaustive circumstances belonging to either are found, reasonable allowances must be made. A maximum consumption, as indicated by experience in Colorado, as well as by comparative estimates in foreign countries with arid climates, may be considered to be, one square inch per acre continually flowing ; and an average consumption to be 72 square inches per 100 acres continually flowing at the rate of four miles per hour, or half a cubic foot per second. This estimate, however, does not include the loss by evaporation, or soakage through the bed of the canal; losses which, in one of the Californian canals, amounts to more than 40 per cent of the quantity entering the mouth of the canal, and is therefore seen to be a very serious item of consideration by the hydraulic engineer. Some further remark upon this important point will be found in the chapter on Canals further on, to which attention is directed.

The art of irrigation, however, is in its infancy with us as yet ; and although we enjoy some special advantages, there are some things to be learned before the full benefit of the practice can be reached. To some extent, our appliances are rude and ineffective, and the watering of crops is sometimes done in such a way as to be injurious to them or wasteful of water. Farming by irrigation, beneath an atmosphere in which evaporation is excessively active, requires special skill to avoid misfortune, and the payment of those costly fees which experience demands when employed as a teacher. To recapitulate some of the most important points to be remembered, might be useful here. The first danger into which the inexperienced irrigator falls, is usually the use of an excessive quantity of water, of a too frequent application of it. The copiousness and frequency of the watering must depend upon the character of the soil and subsoil, to a very great extent. A porous, sandy soil, with a similar subsoil, can hardly be injured by over-watering so long as stagnant water is not allowed to remain upon it, and it is sufficiently well fertilized to bear the vegetation which copious waterings would encourage. Saturation of the soil, long contiuued, would be fatal to almost every crop. A soil containing 80 per cent of sand, may be copiously irrigated every five days without injury, while another containing but 20 per cent of sand, would not bear moderate irrigation more frequently than every 10 or 15 days. The watchful care of the cultivator must be exercised to keep the soil moist and mellow and no more. Over watering tends to bake the soil. Flooding the surface also tends to the same injurious effect. Water should be applied in the evening, in preference to any other time, but on no account in the day during the prevalence of sunshine or a drying wind. A calm evening is the very best time to irrigate. The soil then dries upon the surface before morning, and the sun will not bake or crust the ground. During an occasional shower is a specially favorable time, and this opportunity should be seized and utilized without delay. The use of drills, or small water furrows, is preferable to any other method of applying water. All cultivated crops should therefore be sown or planted in drills. Any crops, that may be grown in ordinary cultivation, may be raised by irrigation, but there are some that flourish better under it than under ordinary culture. These are generally the broad-leafed crops, leguminous plants, and the grasses cultivated for fodder. Long taprooted plants, clover, lucern, carrots, all species of beets, and the cabbage tribe, especially thrive under irrigation. Cereals generally need but very little water after their inflorescence, and the quality of the grain is improved by its absence after fertilization has taken placc. It is asserted by the French irrigators that the wheat crop is frequently injured by any watering at the time of blossoming, and that at this critical season the water shonld be withdrawn, and again applied, for only a very short period, as the grain is swelling. Potatoes are injured in quality by overwatering, and for this crop a soil of retentive character should be specially avoided. The saturation of the subsoil, during the period when the soil is bare of crops, as in the Fall and Winter, not only aids the Summer growth by furnishing a reservoir of moisture, but irrigation at these seasons brings to the soil considerable access of fertility, especially when the water is derived from mountain streams. On the other hand, when the subsoil is strongly alkaline, as in some localities, continuous and copious Winter irrigation will remove much of the excess of alkaline salts ; but alternate irrigation will not have this effect, for much of the alkaline matter will be brought back near the surface by capillary attraction.

There are large tracts of land, the subsoil of which is so thoroughly impregnated with alkali, as to render the surface hopelessly barren, except so far as they may bear a sparse vegetation of plants, the roots of which remain near the surface, and the quality of which unfits them for any use to the stock-raiser or farmer. These lands may easily be reclaimed by irrigation. Copious watering, continuously applied, will wash out the soluble alkali from the subsoil and render them arable. But the watering must be long continued, and at a season when evaporation is the least active. It is the evaporation of moisture from such soils, that brings to the surface the alkaline matter, where it effloresces and makes them appear as if covered with newly fallen snow or hoar frost. All this injurious matter, chiefly consisting of soda salts, may be removed through the subsoil by the continued action of irrigation, and washed into the rivers and the sea. The water charged with these salts, which in excess are destructive, but in moderate supply are helpful, may in fact be used over again on its passage down the rivers after it has emerged beneath the surface in hundreds of springs, upon new fields, which actually need this alkaline matter to make them fruitful.

Another highly important consideration presents itself. It is found that after a few years of irrigation, the soil requires the artificial application of less water; that the atmosphere becomes more highly charged with moisture, and that the evaporation from the surface becomes, in consequence, less and less as years pass; that the rainfall is increased, and that the supply of water becomes relatively more abundant, as the land needs less of it, and thus the area that may be irrigated, gradually increases. The low lands are also moistened by the surplus from the bench lands which percolates through them; the soil becomes charged with vegetable matter, and more retentive of water, and these effects react upon the climate.

These effects have been more particularly noted in Utah, where irrigation has been longest in use, and where the growth of trees has been comparatively extensive. Already the increase in the rainfall has become noticeable, and the level of Great Salt Lake has risen several feet, within the last few years. This, however, can occur only to a limited extent, as the physical features of the country, upon which the peculiarities of the climate depend, must remain permanently as they are, and their effects must of course continue with them. But the intensity of the drouth, and of the hot, dry winds, will probably become ameliorated more and more, as the cultivation of the soil extends.

The management of the various field crops under irrigation, calls for some judgment, and a few general remarks may be useful.

Wheat.—This will always be the main crop wherever irrigation is generally used. A thorough soaking of the soil, some days before it is plowed, is advisable. It then turns up mellow and in fine condition for the seed. Where wheat is made to follow wheat, the seed may be sown upon the stubble, and a light furrow turned over it. Otherwise it would be preferable to drill in the seed, and immediately roll the land with a corrugated roller, fig. 96, which leaves the surface covered with small channels, admirably fitted for watering the crop. (This roller, as well as another kind for the same use, is described more fully in the next chapter.) If no rain should occur, or in localities where rain is not to be looked for, a moderate watering may then be given before the soil has become dried. This will be sufficient to start the growth, after which moderate waterings, at intervals of seven to fourteen days, will be required, up to the time when the grain is heading. Then occurs the critical period, for overwatering may rust the crop ; and it is precisely here that the irrigator enjoys the advantage over the farmer who depends on rainfall exclusively, and frequently sees his hopes and his crop blasted together by unfavorable weather at the period of flowering. It may be that water will be required as soon as the grain is beginning to form, but if the soil is at all moist, water may not be needed.

This point so completely depends upon circumstances, that no rule can be given ; the novice who has never before raised a crop of wheat, will lose less by erring upon the side of caution, and the farmer, used to grow wheat under the ordinary methods, will readily avoid what he knows to be injurious. It will not hurt a crop of wheat if the ground should get dry occasionally, and excess of water encourages growth of straw at the expense of grain.

Other Grains than wheat require very similar management. Oats will thrive with more copious watering, but barley needs care about the time of filling and ripening of the grain. The duration of a watering, for all the small grains, should not exceed 24 hours.

Corn and Broom Corn.—Corn luxuriates beneath heat and moisture ; and for its rapid and healthful growth the soil should be kept moist. The plan adopted in the valley of the Po, in Italy, where maize is a very common and productive crop, is to plant in rows, and apply the water in the spaces between them. The corn may be planted in hills, and watered in a similar manner. As soon as the grain becomes glazed, the water may be withdrawn, and the ground dried for harvesting. Broom corn is managed similarly to maize, being kept regularly watered ; at the time of the heading out of the panicle, water is given plentifully to force a good growth of brush, and produce a smooth, long, straight fiber. The broom corn grown in Tulare Co., Cal., under irrigation, is found to be of the very best quality and color. As these crops require frequent cultivation, the irrigation should be given at a sufficient time before this must be done, to permit the ground to become dry enough for proper working, but not too dry. The cultivation should follow the watering, and not the watering the cultivation ; then the soil is kept mellow and moist during a longer interval. Fodder corn requires copious watering. This crop is one that may be grown to advantage upon fields that are in course of preparation for water meadows, or in rotation, when a meadow needs plowing and reseeding.

Flax.—As this plant, when grown for fiber, depends greatly for its value upon the length and fineness of the staple, and as it flourishes best upon cool, moist soils, it is one peculiarly well adapted for cultivation by irrigation. It may be sown in drills, nine inches apart, or if sown broadcast, the surface should be rolled with the corrugated roller, forming furrows, either directly down, or diagonally across, the slope of the field.

Hemp.—This crop is peculiarly adapted to irrigation, its yield and quality being both improved under this method of cultivation. The mode of culture is as follows. The land is laid off, by the plow, into beds or flat ridges, three feet wide, with intervals between them of one foot in width. The seed is sown upon these beds while the soil is moist from a previous irrigation. The spaces between the beds are left to provide room for hoeing and weeding the beds, and for pulling the male stalks as soon as the pollen has been shed, as well as for irrigation. Hemp is a plant in which the pollen and the seed are produced by different individuals, called respectively male, or staminate, and female, or pistillate plants. As the male plants naturally soon die, long before the others are perfected, it is better to get them out of the way as soon as they have fulfilled their office of fertilizing the flowers of the other sex. After the seed has sprouted, water is given, but only in the spaces between the beds ; these are copiously flowed, so that the moisture may penetrate through every portion of the beds. The crop is irrigated every 10 days, at least, or still more frequently when necessary. The soil should always be kept moist, but at the same time it should not be saturated. Frequent, moderate irrigations, are employed to within fourteen days of the flowering, when the waterings cease. If the irrigation is continued during the flowering, the fer-

tilization of the female flowers is weakened, and the product of seed decreased. The suspension of the watering leaves the spaces between the beds dry, for the passage of the persons who pull out the male plants, which is done to give more room for the ripening of the seed upon those that are left.

Tobacco.—This crop thrives well under irrigation. The method in use where tobacco is largely grown, is to plow the ground to a depth of seven inches, the manure having been previously incorporated with the soil by plowing. The ground is harrowed smoothly and leveled. Ridges are then thrown up, 18 inches apart from each other, and the surface between them is leveled. The beds are then watered by flooding the intervals, and the ground well soaked. As soon as the soil is dry enough, the plants are brought from the seed bed, and set out on the ridge, 18 inches apart; or more, if a large leafed variety is grown. The day following the planting, water is given, and repeated in two or three days. Then an interval of twenty days occurs, in which no water is given, but the soil is hoed or cultivated. Then water is given every 14 days, or if the weather is very dry and warm, and the soil need it, water is turned on every 8 days. Hoeing is done when needed, after the watering. This is continued until the crop is ready for cutting. In every other respect the cultivation is the same as when irrigation is not used. Under irrigation a leaf of remarkably fine texture, and of a mild flavor and color may be grown. Where the climate admits of it, two crops are grown in one year by means of irrigation. This is regularly done in Algeria.

Cotton has been grown in Southern California, under irrigation, with success. It has been found that the peculiar needs of this crop, as regards its growth of stalk and leaf, the formation of bolls, and the season of ripening, are better supplied by irrigation than in any other way. But few crops need so little water as cotton, and by caution in keeping the soil merely moist, and no more, the plant may be prevented from becoming stunted on the one hand, and on the other, the necessity for topping it, to encourage bolling, may be obviated. The method of planting recommended is, to plow high ridges, or beds, 4½ feet wide, in the centre of which a water furrow is made with a small plow. When the soil has been well soaked from these furrows, the earth is thrown into them from each side, a drill is opened above the moistened soil, and the seed sown in it, and covered with the hoe, not more than one inch deep. If the soil has been well moistened, the seed germinates at once, and only one more irrigation is needed to mature the crop, unless on very light and open soil. The soil is plowed in February, and irrigated and planted in March. The usual methods of cultivation and hoeing are practiced.

Lucern or Alfalfa.—Leguminous plants will suffer from as copious irrigation as may be needed for grass or grain crops. Lucern or Alfalfa being one of the leguminosæ must be irrigated with caution, lest the permanence of the crop be endangered. Its long tap roots penetrate deeply, and if much water is given, and the subsoil is at all retentive, they will die and rot, and the crop is but short-lived. The character of the soil should be ascertained before ground that is to be irrigated is sown with lucern. When this is known, the periods and amount of the irrigation may be chosen with accuracy. In Central France, this crop is extensively grown, and yields amazingly under the warm sun and frequent waterings; but in England, lucern does not succeed. It is peculiarly a crop of warm, dry climates, and in California it has been grown with the most satisfactory results, both upon reclaimed "tule" lands, and valley lands. It there requires watering from once a week to once a month, according to the character of the soil. As long as moisture is within reach of the roots, the surface may be left dry, but stagnant water in the subsoil would be fatal to the crop, and must be carefully avoided. Twelve to fifteen tons of fodder have been grown per acre, with four or five cuttings during the growing season, and a watering after each cutting.

Clover is a plant that delights in a cool climate, and where lucern can be produced successfully, it would not be advisable to grow clover under irrigation in competition with it. Under partial irrigation, and where lucern is not a successful crop, it may be watered moderately at intervals of 10 to 14 days, according to the nature of the soil.

Fodder Crops.—Mixed crops of oats and peas; barley and tares, millet, or Hungarian grass, may be grown in succession during the whole year, where frosts do not occur, or during the summer elsewhere. By sowing in drills or forming water channels with the roller, as before described, water may be given with facility during the earlier stages of these crops. When the ground is hidden by the herbage, no further watering is given.

Sorghum.—As a fodder crop this plant cannot compete with corn; but when grown for the manufacture of syrup, it yields largely when irrigated up to a certain point. Its growth is slow and weak at first, and at this stage it will need copious irrigation, so long as the soil is not saturated. Afterwards, when it has commenced its active growth, water should be given sparingly, otherwise the sap will be impaired in quality, No water is given to this crop for a month before cutting, unless from some unexpected cause it is seen to suffer for want of it, and then only the most moderate watering is to be given.

Sugar Beets.—When grown for sugar, this plant needs only moderate irrigation, and at lengthened intervals. The root fibers are very sensitive to excess of moisture, and a watering during one night only, will be all that the plant will safely bear. Excessive growth is not compatible with a yield of rich saccharine juice, and a small solid root is the most profitable. This crop, if it is to be irrigated, is planted in slightly raised beds, between which the water is flowed, so that it does not come in contact with the bulbs. When grown for stock, beets and mangels may be more copiusly watered, until fully grown, when water may be withheld while ripening is completing.

Teasels.—Although this is an uncommon crop, yet as it is grown under irrigation, as a twin crop with winter wheat, it is mentioned here. The manner of its cultivation is, to sow it in alternate rows, or drills, with the wheat, or broadcast mixed with the seed. As soon as the wheat is harvested, the ground is watered, and the irrigation is repeated two or three times, the same season, and monthly the next season, up to a short time before the crop is ready for harvesting.

In concluding this Chapter, it may be as well, at the risk of repetition, to observe, that in irrigation, the object is to supply simply the natural wants of the plants grown upon the land, and not to stimulate an undue or excessive growth, merely because we may suppose that we have the means to do this at our control. The purpose is to supply nutriment to the plants, and not to saturate the soil. The careful irrigator will study the peculiarities of the plants he cultivates, and the character of the soil he works with, as well as something of the natural laws of plant growth ; and apply his knowledge to his business, carefully, systematically, and judiciously; not proceeding in a hap-hazard or a "rule-of-thumb" manner to deluge his soil with water, simply because he has paid for a certain quantity of it.

CHAPTER XVII.

PREPARING THE SURFACE FOR IRRIGATION.

The method of farming by irrigation is very simple and easy to learn. The principals upon which it is managed are summed up in the general laws that water always runs down hill, and that a certain quantity of it is needed for the growth of a plant. In preparing the ground for irrigation, then, it is only necessary to remember these facts and conform the practice to them. The surface of a cultivated field should therefore be of slight slope, generally in one direction, and of an even, smooth character, free from irregularities or knolls. If, however, the character of the surface is such that it is variously inclined with irregular depressions, having a general course downward from the level of the water supply, the courses of the distributing channels may be so laid out as to practicably reduce the whole field to a regular slope and make it very easily irrigated. In the first case, the water taken from the canals of supply will be brought into the main distributing channels, the course of which will be down the slope; directly, if the declivity is not too great, and diagonally if not more than three feet in a hundred. From these channels the water will be taken laterally into other channels, and from them spread over the ground. This plan being suitable only where the soil presents a plane surface, inclined from the canal downward, is obviously fitted for only a very few cases, for those in which the land is altogether free from swells and variations from a level, are very rare naturally and not very common artificially.

Where, however, it is possible so to prepare the land that this even plane surface can be secured, it will manifestly be the best and cheapest in the end, so to prepare it. The great majority of the river bottoms in those parts of the country where cultivation by irrigation is now practiced, and where it is destined to be largely extended, admit of very easy preparation by plowing, harrowing, scraping, and rolling. To plow these lands, a different system from that generally practiced should be adopted. The swivel plow is the best instrument for this purpose. With this plow, the furrows may be laid the same way over a whole field, and the "lands," more or less narrow, necessarily formed with the common plow are avoided. In plowing in "lands" the alternate back furrows and open furrows leave a succession of ridges and hollows, which are inconvenient in irrigated fields, except in those cases in which the system of "bedding" of the soil is adopted. In this case the water is carried along the summits of the lands, and flows in both directions to the open furrows on each side. This may be conveniently done when the general level is once secured.

A good surface will be best secured by using the swivel plow, beginning by running a back furrow across the center of the field, carefully laid out exactly parallel to two of its sides—if the field is square—and equally distant from each. The back furrow should be made by first throwing two furrows outward in opposite directions, leaving an open furrow on the line laid out. The plow is then driven through the center of the ridges thus cast out, splitting them and throwing the earth back into the open furrow. This method leaves no unplowed ground, and very much less ridge in the back furrow than any other manner of beginning the "land." The plowing then proceeds in the usual manner, finishing one side of the field, and then the other. If care is taken to plow straight and even furrows, the last furrow will leave a ditch along the boundary of the field, and close to it. There should be no baulks made in plowing an irrigated field, as the hard spots there left will not absorb water equally with the other portions, and the crop will suffer on those spots. It might be mentioned here that a very smooth, fine surface, is objectionable, as being very liable to bake under the hot sun after watering. A soil that is somewhat cloddy or lumpy, is not so apt to bake, and is preferable to a very fine one. The ground is then leveled with a scraper, the hollows filled with earth from the ridges and swells, and as accurate a level as possible is secured.

When the level surface has been procured, or where it is already sufficiently level naturally, the course of the furrows is to be laid out with due regard to the position of the chief supply canal, and the foot drain by which any surplus of water is to be carried off. It is understood that the chief supply canal is made with as little fall as possible ; in practice, this should not exceed 3 feet per mile, and should not be less than one foot per mile. From this canal the primary, or main distributing ditches are made to diverge, and these should have a slope from 3 to 8 feet per mile ; the medium slope of 4 to 5 feet per mile being preferable. From these primary ditches, secondary ditches are laid out, having the same slope, about 1,000 to 1,500 feet apart, when a large tract is to be brought under irrigation ; a distance of one fourth of a mile, or 1,320 feet, is a very convenient distance, as it is equal to the size of a 40 acre lot, and divides an 80 or 160 acre tract into equal portions. A catch-water ditch should be laid out parallel to the primary ditch, at about 2,000 to 3,000 feet distant from it; a half mile, or 2,640 feet, is a very suitable distance, as there would then be 160 acres, or two 80 acre farms in the enclosed quadrangle.

As an illustration might be represented a plot of an irrigation system, belonging to the San Joaquin and Kings River Canal in California, as described in the report to Congress of the Commission for the examination of the valleys of California. This is shown in the diagram, fig. 81. The main supply canal has a fall of a foot in the mile, the ground under irrigation sloping 8 feet to the mile. The dotted contour lines in the plot represent each foot of slope upon the ground. The water is delivered from the main canal into the primary distributing ditch, at *A*, *D*, flowing in the direction of the arrow. This ditch has a slope of 8 feet to the mile. From these ditches it flows into the secondary ditches, *A B*, *D F*,

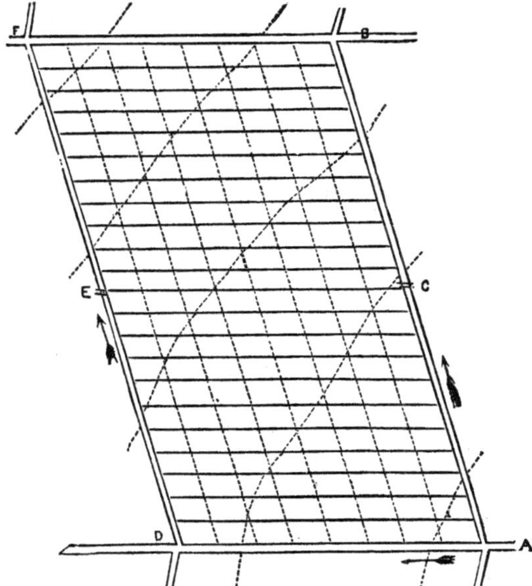

Fig. 81.—PLOT OF IRRIGATED FIELD.

(which have a slope of 5 feet to the mile), through gates at *A*, and *D*, into the catch-water ditches, *B*, *F*, from which it flows into other series of secondary ditches beyond. Gates are established in the secondary ditches, midway of their length, as at *C*, and *E*.

The subdivision of the plot is as follows: Furrows made with the plow or with a ditching machine, which finishes a perfect water furrow at one operation, are run at intervals of 120 feet apart, parallel to the primary ditch, and down the slope of 8 feet to the mile. These are shown by the single dark lines. Other furrows are made parallel to the secondary ditch, and 156 feet apart. These shown by the dotted lines are called check furrows. The secondary ditches are made large enough to supply 11 of these first furrows, each of which communicates with the secondary ditch, by means of a box, such as is shown at fig. 61, (page 128), placed in the bank as seen in the engraving, and opened and shut by a slide at the head of each. When the gate at *C* is closed, the water is turned into 11 of these boxes, and from them into the connected furrows. The first check furrow stops the flow, and dams the water back over the space of 165 feet above it. As the slope of the ground is 8 feet to the mile, the slope of the interval now covered with water is nearly 3 inches, and the water must consequently be 3 inches deep at the check furrow before the upper portion of the interval is watered. (Here a fault in the lay-out is seen at first sight, because from the rapid absorption of the water by the soil, either the lower portion must be watered to excess, or the upper portion be left without a sufficient supply. It is evident that this fault would be obviated by making the check furrows nearer together, say 50 or 60 feet, when the ground would be more quickly covered, and more evenly watered. It is true that some of the water would soak through the check furrow on to the upper portion of the interval below it ; but this would be an irregular and entirely a too hazardous proceeding to be adopted by a careful irrigator, and one that would be excessively wasteful, both of water and crop.) When the interval has been watered sufficiently, the check furrow is opened with the hoe at each main furrow, and the second strip is watered. The process is repeated until this half of the plot has been watered. The boxes are then closed ; the gate at *C* is opened, and the other part of the plot is irrigated in the same manner. The size required for the secondary ditch, or rather that of the gate at *A*, by which the ditch is supplied, must be proportioned to the quantity of land irrigated by it. If the plot is 160 acres, the gate should have an area of at least 144 square inches, if the flow is continual, and a proportionately larger area if the flow is intermittent. The size of the boxes should be in proportion, or $14\frac{1}{2}$ square inches for each inside measurement. A box $7\frac{1}{4}$ inches wide by 5 inches deep inside measure, and having

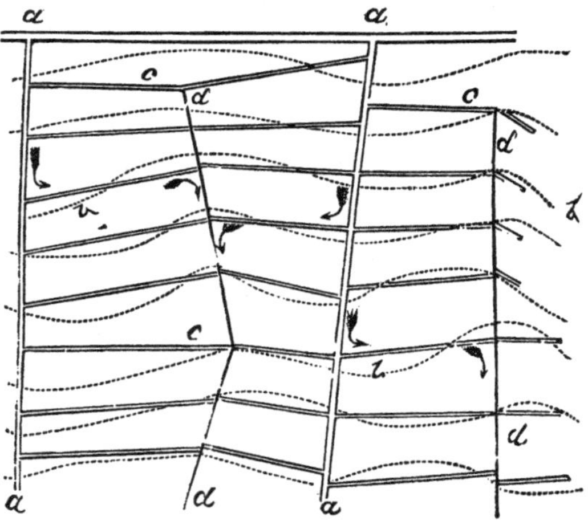

Fig. 82.—PLAN OF FURROWS FOR AN IRREGULAR SURFACE.

a gate or slide to open 2 inches, would give $14\frac{1}{2}$ square inches of water under a head of 3 inches, which is a usual arrangement for supply.

This plan is an excellent one, and pointed out as an illustration of ordinary irrigation, could hardly be excelled. With modifications, it offers a method of preparing the surface of gently sloping ground that is applicable to a wide diversity of instances.

The plot given in the illustration is drawn to scale of

960 feet to an inch, and represents a plot of 80 acres, being 2,640 feet, (half a mile), in length, having 22 furrows, 120 feet apart, in that direction; and 1,320 feet, (quarter of a mile), in width, and having 8 check furrows, 165 feet apart, in this direction.

For those cases in which an irregular surface cannot be avoided the arrangement of the water furrows is different. For a field which slopes on either side from a central ridge the arrangement is made as follows: (See fig. 82.)

The water is brought from the main primary ditch, on to the highest portion of the ridge. From this it is carried by a principal furrow along the ridge, and then by other furrows, *a, a,* on each side down the slope, and from those into distributing furrows, *c, c,* nearly parallel with the main furrow, and in the manner before described for the bedding system applied to gardens. Thus the water flows down the slope on each side in a series of channels provided for it, according to the circumstances or necessities of each case.

Where the surface is irregular in every direction, it is necessary to discover by careful leveling the course to be taken by each main distributing canal, which should be made the boundary line between the fields on either hand, and will therefore be a permanent construction. The course of the canal will be such as to give it the least possible slope, so that none of the head of water may be lost in distributing it from the end of the canal. It is obvious that this course should be laid out with great care and exactness, lest by losing some of the head, some portion of the land should be left without water. This work should be done either by a competent surveyor, or by the assistance of instruments in the hands of the farmer competent to use them. There is no particular skill required to do this; it is rather a work that calls for extreme care and patient verification. The instrument used may be a surveyor's field or portable level, which will answer every purpose for light and not very accurate work. A very portable and convenient level, small enough to be carried in the coat pocket, has been found by the author of very great use in making preliminary surveys. It is known as Locke's hand-level, and is shown in fig. 83. Very accurate levels may be taken by using this instrument in the following manner. A rod is provided, having a blunt foot, that will rest upon the ground, and not sink in soft soil, and of such a graduated length that it will reach comfortably to a hight equal to that of the eye of the person using it. The top of this resting rod is slightly notched, so that the level will rest easily upon it. By having the sighting rod marked at exactly the same length from the foot as that of the resting rod, and gauged up and down from this mark, (which should be an

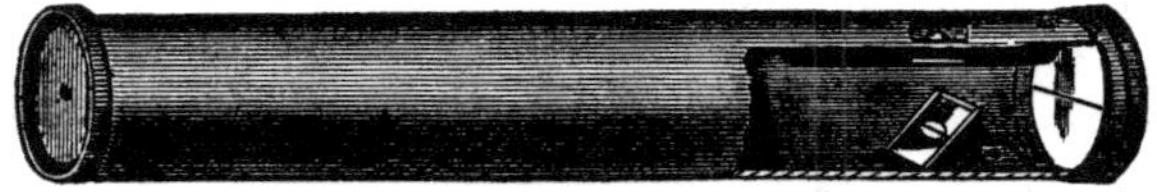

Fig. 83.—LOCKE'S HAND-LEVEL.

0), the variations from the level may be taken with the greatest readiness, and with sufficient accuracy for preliminary work, or for a survey, where complete exactitude is not required. Any slight errors that may be made will balance each other, and in the aggregate there will be very little variation from a true level in a line of some miles in length. In the illustration the side of the level is represented as broken away, to show the mirror in the interior which reflects the bubble and the cross-bar in the center. The bubble is seen at the top of the level.

The Architect's level, fig. 84, made by W. & L. E. Gurley, of Troy, N. Y., is a more costly and complete level, but a very simple, compact, and serviceable one. It has a telescope 11 inches long, with the usual cross wires, with adjustment of eye and object tubes. It may be mounted on a Jacob-staff, or a tripod; but for sighting along such work as irrigating canals and embankments it may be placed upon a three cornered plate of iron, or a trivet, standing upon three pins, by which it may be firm-

Fig. 84.—GURLEY'S LEVEL.

ly set upon a piece of wood, earth, or a stone. A tripod and the trivet is furnished with the level by the makers, for the very reasonable price of $35. A leveling rod with target is $5 extra.

An instrument much used by French irrigators is thus made: Two pieces of wood, 1½ inch wide by 1 inch thick and 10 feet long, are pinned together at one end, forming an angle, the base of which spreads exactly 16½ feet. This gives a hight of the apex or joint from the ground of about 5½ feet. The arms are fixed in their proper position by a cross-piece at about 3½ feet from the base, and fixed exactly parallel to it. The end of each arm is

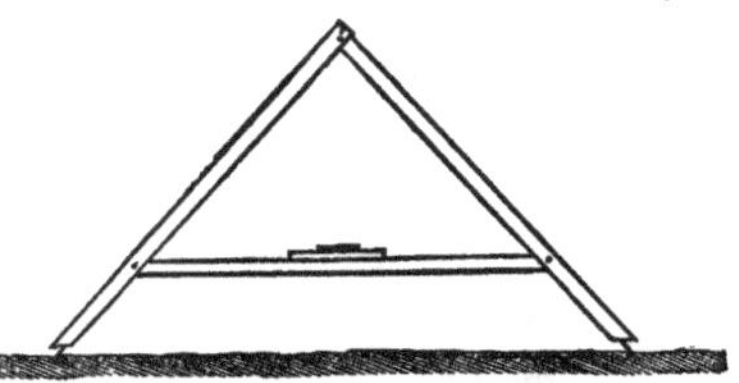

Fig. 85.—HOME-MADE LEVEL.

pointed and protected by a metal ring or ferule, and a pointed iron pin is inserted. The implement is something like a large pair of compasses with a spread of 16½ feet between the points. See fig. 85. This distance is not arbitrary, but may be varied to 10, 12, or even less feet, but more would be inconvenient. But 16½ feet being exactly one rod, the level may at other times answer for a measurer of distances conveniently, if made of this size. A spirit level is placed on the cross-bar, see fig. 86, care being taken to place it exactly parallel to the line

Fig. 86.—ARRANGEMENT OF THE SPIRIT LEVEL.

between the bottom pins, and to verify the parallelism by reversing the position of the implement as it stands upon the ground upon a spot shown to be level by its first position. There is no difficulty in getting the level exactly placed to one who understands the use of the spirit level, but unless it is placed exactly the implement will be useless. To use the implement, a wooden plug is driven in the ground, level with the surface, at a point where the canal is to start from. One of the legs of the level is placed upon the plug and the other forward in the direction in which the water is to flow, and from one side to another, until a point is found which is level with the starting point. A plug is then driven into the ground at the second point. If the fall has been fixed at one foot in 1,000 feet, which is the most advisable for distributing furrows or canals, the second plug should be placed a fifth of an inch below the level of the first. This may easily be done with accuracy by cutting off with a saw, one-half of the top of the plug one-fifth of an inch below the other half, when they are prepared for use. Figure 87. Then when the higher

Fig. 87.

half is placed on a level with the lower half of the preceding plug, the lower half will be exactly in position to receive the leg of the level to lay out the next space. In this way the ground is gone over until the whole line is laid out and the end is reached. From these trial contour lines furrows may be traced and laid out, using pegs, in the same manner as before, and a plow to make furrows, following the line of the pegs. Or one man using the level nearly as fast as he can walk, may be followed by a boy or another man with a hoe made for this purpose, with a blade 18 inches in width, with which a furrow is rapidly opened, almost as fast as the line can be laid out. The distributing furrows may be laid out in straight lines across the contour lines, see fig. 82, which will save labor. The main canal, *a, a,* fig. 82, passes along the highest part of the field. The contour lines, which are the lines of level, see dotted lines *b, b,* are run from the canal on either side and meander with the irregularities of the surface. To avoid the meandering of the distributing furrows, they are made to run in straight lines from the canal, cutting across the contour lines with what fall may be found necessary ; these furrows are shown at *c, c, c.* By regulating the supply of water so that all is absorbed, none need go to waste. But it would be safe to run a drainage furrow to carry off any accidental surplus across the lowest portions of the distributing canals, as shown by the dark lines, *d, d.*

When the furrows are properly leveled, the soil may be watered, either by saturation from the furrows downwards, in the case of steep hill sides, or by tapping the furrows, and causing the water to escape downwards from them. The method of watering lands of considerable slope ; that is, of more than five feet in a hundred, or ten inches in 16 feet, must be different from that previously described in this chapter, or by flooding ; on the contrary the water must be led downwards from the furrow in a thin sheet or in numerous trickling streams, which may be made to cover the intervals between the furrows. There will be, however, some instances, and in time, after the best of the irrigable lands have been occupied, the majority of the tracts left will be of this character, in which the surface will offer more than usual difficulties in the way of preparation for irrigation. These tracts referred to are hilly lands, or so called foot hills; high prairie lands or bluffs bordering the more tractable river

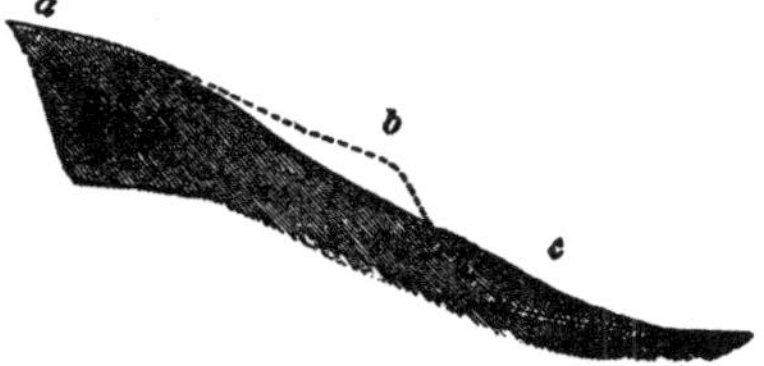

Fig. 88.—IMPROVEMENT OF A HILL-SIDE.

bottoms and valleys. The surfaces of such lands are in general cut up with hollows, ravines, gulleys, and similar irregularities of a somewhat miniature character, but which nevertheless offer obstacles to the passage of water channels; or there may be aprupt declines and rounded protuberances, which will require modifying to some extent. By some system of preparation all such lands may be brought under irrigation, and a few typical cases are

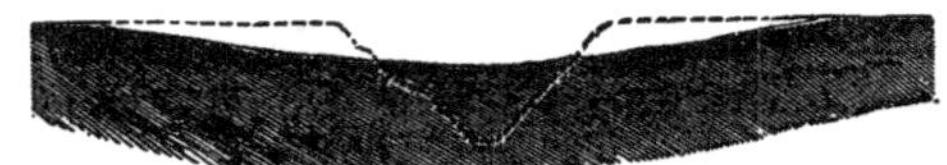

Fig. 89.—MANNER OF FILLING A GULLEY.

here referred to with the requisite treatment. A profile of a hill too steep in one portion to be irrigated, is represented by the dotted lines in fig. 88. The rounded outline, *a*, *b*, *c*, offers an obstruction to both water furrows and the passage of men or animals. By cutting away the projecting portion at *b*, and depositing it in the bottom at *c*, the outline, as shown by the dark line, is made passable and easy to irrigate by the method applicable to lands of considerable slope, (see figures 73 and 74 with accompanying descriptions), or by that described in the preceding paragraph. A gulley or a hollow in a moderately sloping surface is shown by the dotted line in fig. 89. This difficulty is removed by taking away the portions above the dark line, and depositing them in the hollow beneath it; thus bringing the new surface into conformity with that surrounding it, and producing an easy slope. In case the surface soil is thin and the subsoil poor, it will be necessary to first remove the surface soil from both the portion to be covered, and that to be moved, and place it on one side. When the leveling is finished, the

Fig. 90.—TERRACING A HILL-SIDE.

surface soil is returned and the subsoil covered with it as before. There are frequent hill-sides, all through the country, which offer no impediment to a destructive flow of water down their slopes, and unsightly and inconvenient gulleys and wash-outs are caused by this unobstructed flow. The terracing of such slopes would prevent the destructive wasting, and would render them amenable to easy irrigation, either by surplus rain water collected from the slopes in reservoirs, or by water brought to them by elevation or otherwise. Fig. 90 is intended to represent such a hillside. The original profile is shown by the dotted line, and the terraced outline by the dark line. This work may be done almost wholly by the plow, and in difficult cases partly by the plow, and partly by the ordinary horse-shovel. Upon the remodelled surface the water is retained, instead of flowing in streams uselessly and destructively down the slope, and sinks into the soil moistening the whole as it percolates through the subsoil and again reaches the light, lower down. Or the terraces may be so arranged as to lead the rain water into a reservoir, where it may be stored, and used to irrigate the lower portion of the slope in the drier part of the season.

As the preservation of a level or smoothly sloping surface is the main point in preparing the soil for irrigation, it is important to have implements well adapted to this necessary work, and also to prepare furrows quickly and perfectly. There is no need for costly implements, but very effective ones may be constructed with little labor and skill. To level the ground is the first work after plowing and pulverizing the surface. To do this cheaply, a scraper that can be operated by horse-power is needed. One upon which the operator can ride would be most convenient, as the work may then be overlooked with ease, and the weight of the rider would add to the effectiveness of the implement. A horse-scraper, much used in California for leveling ground plowed for irrigation, consists of a frame, 4 feet wide and 6 feet long, mounted upon a pair of low wheels, and constructed of planks, upon which the driver rides. A tongue is fixed to the central part of the frame, by which the machine is drawn along. A scraper is fixed to the front of the frame in a perpendicular or a sloping direction, as may be desired. Handles, or guides, are fixed to the scraper, by which this direction is governed. The scraper is a plank, 12 feet long and a foot and a half wide, shod at the bottom edge by a steel shoe. A half circular, flat, iron bar is bolted to the front of the scraper and passes through an iron strap fixed beneath the tongue. The bar is pierced with a number of holes, and a hole is made through the tongue so that an iron pin may be passed through both tongue and bar. By means of this contrivance the plank may be moved from a straight to a diagonal direction across the path traveled, and the earth is consequently drawn forward or thrown

Fig. 91.—EARTH-SCRAPER.

off to one side, or both together. In this way a newly plowed field is leveled very quickly, and is easily prepared for furrowing. A scraper very easily made is shown at

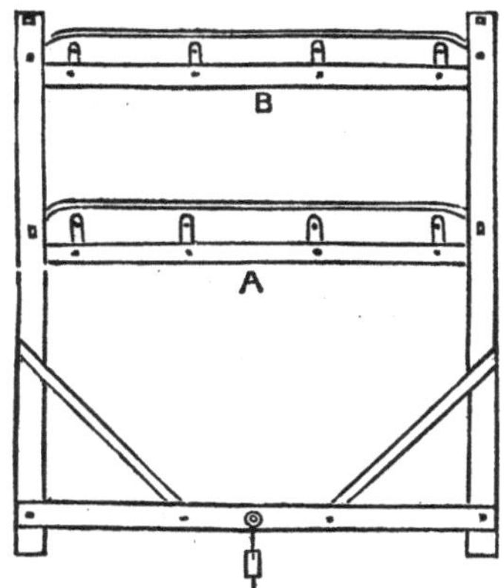

Fig. 92.—FRENCH SCRAPER.

fig. 91. It consists of a central plank and two other planks hinged to it as wings, and adjusted in different positions, and so held by means of strong braces. It is shod with steel, and is furnished with a tongue for draft. By adjusting the wings the earth may be scraped in different ways, as may be desired; and ridges may be formed by it, by proper adjustment of the wings and shape of the central plank. Another implement for this purpose is used among the French and Italian irrigators, which is very effective, and is employed as frequently as

Fig. 93.—FORM OF SCRAPER.

the plow. It consists of a frame, seen at fig. 92, of timber bolted or mortised together, and braced with two diagonal braces at the front. It is generally square in shape and admits of being made of any suitable size. Two cross-pieces, *A*, and *B*, are provided with metal shoes, similar in shape to plane-irons, which project beneath the surface, as shown at figs. 93, and 94. As the machine is drawn across the field the scrapers take off every protuberance, and deposit the loosened soil in the hollows, and in time, by passing across the field in different directions, a perfect level is gained. To enable this machine to be transported from place to place more readily, the upper side of the side-pieces may be provided with shoes made of light bar-iron, affixed in the manner shown at fig. 95. When it is to be moved from the field it is simply turned over, and glides over the soil upon these shoes. As the implement will be in constant use it should be stoutly made and carefully preserved. When a smooth, level surface has been obtained, the seed

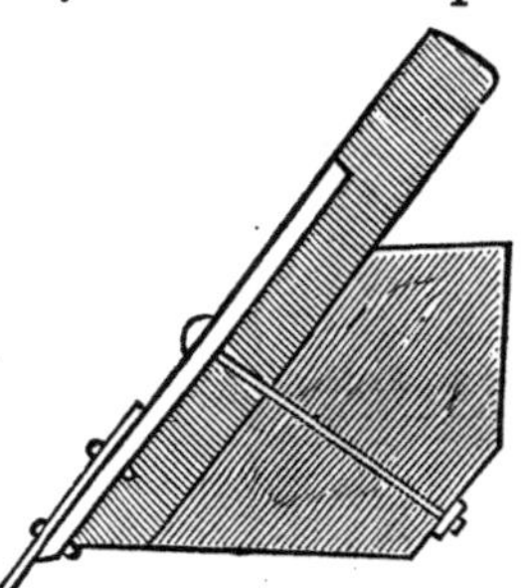
Fig. 94.—ENLARGED VIEW.

sown and the field harrowed, the soil may be furrowed by passing over it in the direction in which the water is to flow upon it, a roller provided with corrugations upon its surface, each of which leaves a small distributing furrow. See fig. 96. This roller may be made of cast iron disks, 18 inches or more in diameter, and of such a thick-

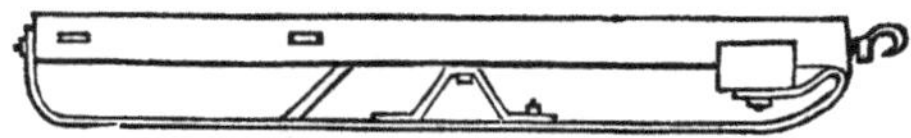

Fig. 95.—SCRAPER INVERTED.

ness as may conform to the distance between the furrows, or the disks may be made of sand and cement, forming in reality artificial stone. The cement may be shaped in wooden molds. These disks will have holes two inches in diameter through their centers, through which an axle, consisting of rolled-iron bar or shaft, may be placed. The axle may be fixed at the ends in a wooden frame, provided with a tongue for draft. By such a method of construction sufficient weight may be secured to compact the soil and make the furrows durable. Another form of

Fig. 96.—CORRUGATED ROLLER.

roller is shown at fig. 97. This may be made of circular sections of oak plank, 30 inches in diameter, with others placed alternately with these, of 36 inches in diameter. These sections may all be independent of each other, but it will be more convenient if they are in pairs or triplets; for the reason that it will be necessary to make these sections of several pieces, and it will be easy to bolt them together by crossing the pieces of one section upon those of another, or two more. The most desirable plan will probably be to make them in triplets in the manner shown

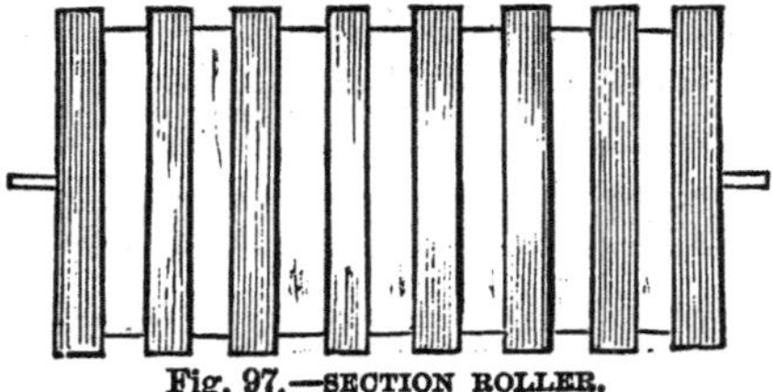

Fig. 97.—SECTION ROLLER.

in fig. 98, the dotted lines showing the manner in which the joints of each section cross those of the others. These sections may be placed upon an axle, as previously described, and provided with a frame, upon which there may be a seat for the driver. Various other forms of rollers may be devised which will answer the purpose of making furrows for those crops that cover the ground entirely, and which are sown either in narrow drills or broadcast. For such crops it might be desirable to make the drills, as well as the distributing furrow, to run in an east and west direction when this is practicable. The ground will thus be shaded from the southern sun by the growing crop, and the moistened furrows will be protected from a too rapid evaporation. The furrows may be made to traverse the ground between the drills, leaving the drills and furrows alternate.

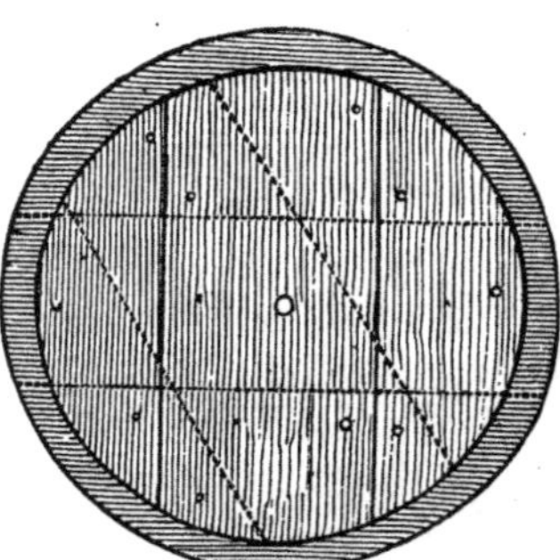

Fig. 98.—FORM OF SECTION.

It is obvious that the use of the rollers here suggested is only applicable to the grain crops, and not for those that are to be cultivated.

CHAPTER XVIII.

THE SUPPLY OF WATER.—DAMS.—PUMPS.—RESERVOIRS.—ARTESIAN WELLS.

DAMS.—For extensive irrigation, the available supply of water can be found only in permanent streams; large and copious wells, from which the water is raised by pumps of great capacity, operated by steam, or in extensive reservoirs, in which the drainage of large areas of mountain territory is collected. No dependence can be placed upon artesian wells, though the contrary has been erroneously taught by some writers having a limited acquaintance with this subject. This expectation has been shown in a previous chapter, to be delusive, both on account of the limited supply of water that can be thus obtained, and the costliness of the system. For exceptional cases, these wells may be employed with profit. These cases will be found to exist where extensive water-bearing strata are depressed in a basin shaped area, at a moderate depth beneath the surface, so that a copious and permanent supply can be procured at a moderate cost, and where the area to be irrigated is small. The futility of depending upon artesian wells, in other cases than those above cited, will be evident when the principle upon which they operate, is explained further on in this chapter.

For the present, and for many years to come the main

supply for irrigation will be derived from streams. The water, in most cases, will be taken directly from the stream at its regular level, by means of a main supply canal, into which it is diverted by the ordinary flow, or by means of wing drains placed in the stream; or else the level of the stream must be raised by a dam, and the flow diverted at a higher level than the usual one. It is always advisable, in fact necessary when profit is the main purpose, to choose such a location for the commencement of the canal as shall give the greatest possible head of water. The cost of a few miles of canal may be insignificant, as compared with the value of several thousand acres of land that may be brought under irrigation, by adding a foot to the head of the supply. But a dam may often be constructed at a much less cost than would be

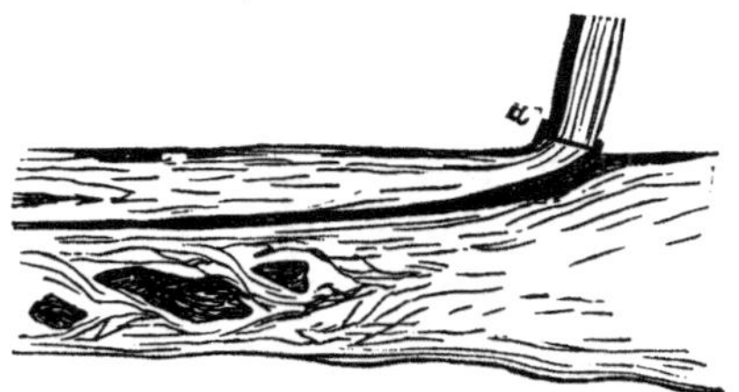

Fig. 99.—LONGITUDINAL WING DAM.

necessary to carry a large canal a distance of even a thousand feet. If the level needs to be raised but a few feet, a wing dam may be constructed. This should be placed where the level of the stream falls sufficiently, and should be carried up the stream, at a convenient distance from the bank, as far as may be necessary to raise the water to the hight required.

The manner of constructing the wing dam will vary according to the character of the stream, the nature of the river bed, and the materials to be most conveniently procured. The typical form of the dam is shown at fig. 99, in which the structure is seen projected up the stream to a point at which the required level is reached. This point should be found by careful survey, before any work is done, because the strength and size of the dam must be proportioned to the pressure of the water contained within it, and this is in a ratio with the hight of head of the confined portion of the stream. This kind of dam very rarely requires elaborate construction, but as it is exposed to frequent erosive washing by the stream, in floods, it should be built of such materials as are known to bind well together. When but little head is required, a very simple dam of brush, stone, and earth, will be sufficient. The work is commenced at the head of the canal, which is first excavated to the proper depth, up to the river bank where the head-gate is, (see *a* in the figure), properly constructed. The building of the wing dam is carried up the stream from this point. A few piles driven into the river bed, three feet apart in a double row would be advisable at this point, and for such a distance up the stream as may seem proper, as this point of the dam is exposed to the greatest pressure, and is generally the weakest; for the reason that in all earth-work the junction of the old and new material is the most difficult to consolidate evenly. If brush is to be conveniently procured, it may be interwoven between the piles of each row, and rammed down compactly. Cross-ties may be bolted or pinned to the piles, to prevent them from spreading, and earth is then thrown between them. Brush should be placed between the rows of piles, and the brush should be placed so that the buts lie down stream, and the fine part in the contrary direction. As the earth is dumped into the dam it will cover the brush. Afterwards coarse gravel or stone may be used to fill on the outside. In this manner the dam is carried onwards to the extremity where brush covered with earth will be sufficient to divert the current where the difference between the levels is but slight.

Where a more substantial construction is needed, or where the current is so swift that loose earth would be carried away, the method of construction will be different. Crib work would then be required, or else the piling should be continued to the end, and should be of a substantial character. The piling, or the cribs, should be connected by stringers and cross-ties, and the vacancies may be filled with brush and stone. It is not always necessary that the dam be absolutely tight, as if it were one of the usual kind ; if it diverts a sufficient quantity of water, that is all that is required. But a tight dam may be made in this manner, if the cribs, or the space between the piles are filled with brush and stone, and earth be thrown upon the inner side of the dam. Then

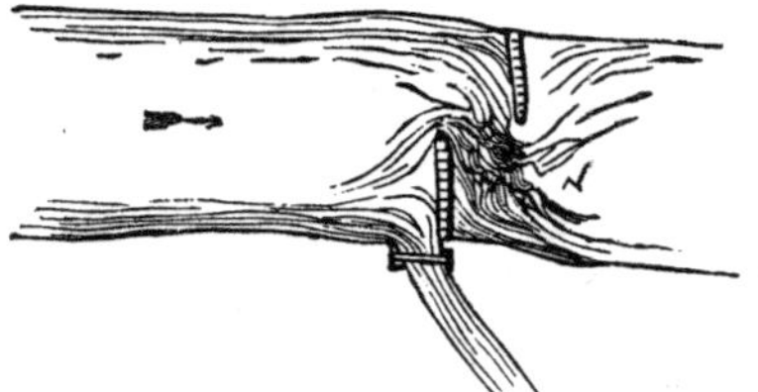

Fig. 100.—CROSS-WING DAM.

the cribs, or piling, serve as supports to the dam, and the earth serves to confine the water. In some cases wing dams of a different form may be used. Where, for instance, a longitudinal dam would need to be of extreme length, because of the inadequate fall of the stream, and where it is desirable to avoid closing the stream entirely, cross-wing dams may be constructed in the manner shown at fig. 100. Here partial dams of crib work, or piles, filled in with stone ; or dams of logs, brush, and earth, are thrown into the stream, from each side, but not upon the same line, so that when each reaches the middle, an open space it left through which a portion of the water escapes in a rapid. The distance between the ends of the dams requires to be arranged so as to raise the level of the water above them to a sufficient hight, and yet leave an open passage with a current that may not be insurmountable to vessels or boats navigating the stream.

When dams of the ordinary construction are required, it may be necessary to consider, before the work is begun, the principles upon which their stability is founded. This is more especially necessary, when the work is of any considerable magnitude, and where a failure may involve the loss of the money spent, and much direct and indirect damage besides. The chief points for consideration in this regard are, the position of the dam in the stream; the material of which it is to be made; the form most consistent with permanence and stability; and the manner of its construction.

The position of the dam has reference only to its power of resisting the pressure of the water behind it. No increase of the flow of water into the canal, or lateral, can be gained by placing a dam in a diagonal position across the stream, instead of at right angles to the banks, as has been stated by some who have written upon this subject in the public journals. As an instance of the incorrect and misleading notions thus spread abroad, by uninformed writers, might be cited the following from an article on "Practical Irrigation in Colorado," published in the Report of the Department of Agriculture for 1871. The writer says, "it has been contended that the stagnation of water extends to a sensible hight, above the horizontal line of the regurgitation from the dam or sluice, or any other fixed obstacle. *This is accounted for by the compression or closer adhesion of the particles of the water.*" Again he says, "if you confine the water, and divert it from its natural course, *you may compress it into a smaller space;* but the same quantity will be found below the compression, as is found above it!" Now, it ought to be known that water is practically incompressible, and rather than submit to pressure, its particles may be forced through the infinitely small pores of cast iron, if the iron is strong enough to resist the enormous pressure required. So many instances of this property of water are supposed to be popularly known, that a statement to the contrary not only misleads and confuses those who read it, but tends to cast doubt and suspicion upon whatever else the writer may say.

To expect, therefore, that by the use of diverging entrances to a canal, or by the use of a funnel-shaped sluice, a larger quantity of water may be forced into a channel, will be found fallacious, and will lead to disappointment. A funnel-shaped box will pass no more water through it, than can be passed through another with straight sides, and of the same diameter as the narrow throat of the funnel, unless the inclination is changed and the velocity increased. This is an established principle of hydraulics. Other principles of hydraulics, which relate to the construction and use of dams, are, that the pressure of water is equal in all directions; that it is exerted only in proportion to the hight and area of the base of the column of water resting upon a given space; that water will always seek and maintain an exact level, and that the disturbance of the level sets it into immediate motion.

The pressure of a body of water upon a perpendicular wall, a dam, or any other obstacle to its flow, is exerted to force it forwards in the direction of the stream. A dam placed directly across the stream is, therefore, weak and faulty. It will be rendered very much stronger by being placed across the stream in a curved, or angular, form, with its apex towards the resistance, and giving it somewhat the shape, and consequent strength, of an arch. The material of the dam should be selected for its impermeability to water, and for its more perfect capacity for binding together, and resisting disintegration. There is no better material for a dam than earth which contains a large proportion of clay, with enough sand intimately mixed in the mass to make it easily worked, and closely compacted. But dams may be made of rock and timber, as well as of earth, the former materials being selected, when the work needs to be of the most substantial character, to enable it to resist the wearing action of strong and heavy currents of water which would tear away an earth work in a short time. Where this contingency is likely to occur, only timber or rock should be used, and the manner of construction should be left to the direction of a practical engineer. For a work of timber, cribbing filled in and backed with rock, and planked thoroughly well, will be found very substantial and satisfactory. There are many different kinds and forms of cribs with which the hydraulic engineer is familiar, of which those may be selected that will meet the particular features of the cases requiring them, and which for want of space cannot be referred to here. A few will be described further on, of those only which may be found useful to the irrigator who desires to perform his own engineering, and in cases where professional assistance may not be required.

Upon the form of the dam will depend, in a very great measure, its strength and stability, for it is evident that the form has much to do with its power of resisting the enormous pressure bearing upon it, and which is always exerted either to overthrow it or to push it from its foundation. Further than this, the form of a dam should be such as will best resist the wearing and abrading action of the water. A typical form of a perfect dam is shown at fig. 101. The reasons why such a form is best adapted for its purpose may be briefly stated as follows:

It is evident that a structure, intended to sustain a pressure of a body of water, can fail only in two ways, if its solidity is preserved from disintegration by the wearing actions of currents. These are—either by being overturned by the horizontal pressure of the water, or by being forced from its position bodily by sliding upon its base. The first alternative may be examined by considering what power a certain structure—a vertical wall for instance—exercises to resist the pressure of water, and what the pressure amounts to for a certain hight. The pressure of water upon any surface immersed in it, is equal to the area of the surface multiplied by the depth of its center of gravity below the level of the water, and by the weight of a unit of water. The unit adopted in these calculations is a foot, and a cubic foot of water weighs $62^1|_2$ pounds. The resulting pressure is therefore

Fig. 101.—GENERAL FORM OF DAM.

readily found. Let it be supposed that a wall 10 feet high is sustaining a body of water behind it, as shown in fig. 102. One foot in length of the wall is taken as a basis for the calculation. There is then 10 square feet subject to pressure; the depth of the center of gravity is 5 feet; and the weight of a foot of water is $62^1|_2$ pounds. The product of these numbers is 3,125, which is the number of pounds pressing upon one foot in length of the wall. But this pressure, in this case, is not evenly distributed over the whole wall, but in consequence of the mobility of the water, the pressure is so distributed as to be equal to, and to operate as, a single force acting at a point one-third of the hight of the wall from the bottom. For this reason the product previously arrived at should be multiplied by one-third of the hight, or $3^1|_3$, which will give as the total pressure exerted to overthrow or push forward the wall, 10,406 pounds on every foot in length. To resist this, there is nothing but the weight of the wall, and as we have already the length and hight, the thickness only is needed to give the required resistance. The rule for finding this, or to be more precise, for finding the required weight of the wall for its stability, is to multiply together the hight of the wall in feet, by half the thickness, and by 112, the weight in pounds of a cubic foot of masonry, and divide the amount of pressure, previously ascertained, (10,406), by the sum given. In this case we get $4\frac{1}{3}$ feet as the required thickness of the structure.

It is evident that this supposed case is one of the weakest illustrations that could be chosen, because a wall of

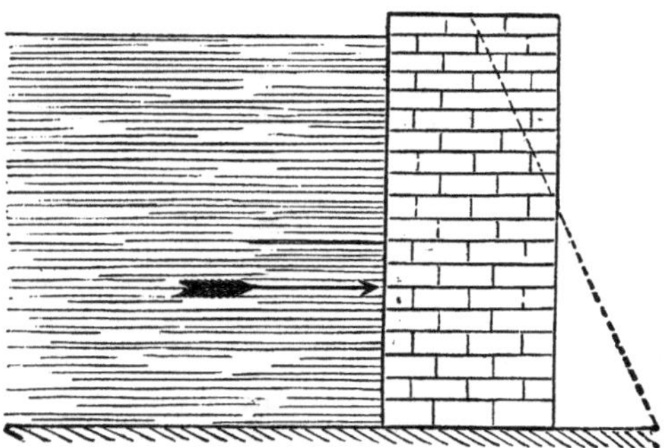

Fig. 102.

this character is poorly calculated to resist the pressure. But it is a perfectly safe method of calculation, because all the errors are on the right side. If we take off a portion of the upper part of the wall, and place it at the bottom, as shown by the dotted line in the illustration, fig. 102, it is clear that we remove some weight from a point where it is not needed, and put it where it will give much greater resistance, both to oversetting, and displacing; removing the point upon which the wall must turn in case of overthrow, and therefore increasing the leverage, and consequent resistance, and also greatly adding to the tendency to resist sliding. The more this weight at the bottom is increased, the stronger, therefore, is the dam. This principle of calculation applied to a bank of earth, or any other construction of the form, shown in fig. 101, will easily show that the power of resistance to overthrow is immensely increased when long slopes are made instead of vertical walls. Besides this increase, the downward pressure of the body of water upon the inner slope, adds to the resistance against both overturn and sliding, and when the foundation is excavated, as shown in the illustration, this tendency to resist sliding is again increased, because the adhesion between the old and new earth is rendered more perfect. The thorough incorporation of the old and new surfaces of earth must be carefully made, as a preliminary condition of stability. The full conditions of stability include a weight of bank which with the vertical pressure exercised by the water, to hold it down, will equal the horizontal pressure of the water, against the dam, and leave a surplus to meet any unexpected contingencies. In addition to these, the materials of the construction must be of such a character as will resist percolation of the water, and will bond together intimately and with cohesion. It is not often that dams give way by sliding upon their foundations; but an instance of this has happened in the author's experience, when from faulty construction an earth dam, founded upon a smooth rock bottom, gave way bodily to the pressure of the water. But this dam was made by an inexperienced man, in defiance of professional advice, and of proper principles of construction.

The best examples of the inside slope of a dam is either 3 feet horizontal to 1 foot perpendicular, or $2\frac{1}{2}$ to 1. The outside slope may be from $1\frac{1}{2}$ to 3, to 1, depending upon the character of the material, and the means used to prevent the surface from washing or crumbling away.

These may be either by covering the face with sods, in case no overflow is permitted, or with masonry or planking.

The manner of constructing a dam is of the greatest importance. The modern method is to introduce a puddle wall in the middle, to place selected materials upon each side of this, and to form the slopes of the most convenient materials to be procured, whether gravel, rubble stone, or waste broken rock. But there are many very ancient embankments, still existing, that have been constructed without puddled centers, or any special precautions to make them water-tight. The ancient manner of making these embankments was, to carry the earth in baskets upon the heads of the workmen, and deposit it where it was required, without any particular care as to the disposal of it. The constant treading and the thorough consolidation of the earth, by being thus thrown in small quantities beneath the feet of the workmen, tended to make a well incorporated, homogeneous mass, which would be impenetrable by the water. It would be difficult to discover any better mode of construction than this. A dam constructed by the author upon an uneven rock bottom which furnished an excellent foundation, and of a crumbly, loamy clay earth, which melted down in water to a pasty mass, was made without any puddling, and by simply carting the earth and dumping it into its place ; the stream having been previously confined within a flume of timber where the waste gate was afterwards put in. The treading of the horses and men, and the packing caused by the cart wheels, so perfectly consolidated the earth that no leak was observable, and the dam is now probably better than it was when first made, 15 years ago. This dam was faced upon both sides with waste broken rock, and in one severe freshet, water has poured over the top to a depth of more than two feet, for several days, without any injury.

As a general rule, for dams of not more than 20 feet in hight, when earth of the best kind, or such as is mentioned above, can be procured, puddling may be dispensed with. When puddling is used, it would seem to be more properly placed upon the inner side of the work, with the selected material next to it, and the poorest used as a backing to support the work ; this, although seeming reasonable, is not in accordance with practice, and no one seems inclined to risk the innovation upon an accepted custom, with the risk of blame for it in case of failure from whatever cause.

The first requisite, in constructing a dam of any magnitude, is to ascertain the character of the foundation, and to excavate this to a bed of solid rock or impermeable earth. If springs are encountered the location may be abandoned and another chosen, or else the spring must be carried away in tight drains, beyond the outer slope of the dam. The channel for the water flow is then to be constructed—in case the dam is to be used for a reservoir—in the solid subsoil, and the pipes or culverts used for this purpose should be flanged every few feet of the length, that the puddling around them may be more thoroughly compacted, and the danger of leakage at this most important point, by the creeping of the water along the surface, be prevented. All disturbance to the pipes or culverts, by settling of the work, which might occur if they were placed in the body of the dam, is thus avoided.

The best of the selected material is then disposed in thin layers upon the foundation, and well rammed, or puddled. The puddle wall may be carried up in the centre of the selected earth or clay, or upon the inner face of it; which, although an innovation upon established practice, would be an improvement upon it. The earth should be brought to the dam by carts, in preference to wheel-barrows or to tip-cars upon a track although the expense may be greater, because of the more perfect consolidation of the work by the trampling of the horses, and the cutting of the wheels. It should be disposed in regular layers, of 2 to 3 feet in thickness, over the whole work. These layers should be depressed at the center of the work, so as to give a basin shaped form to the section. This is shown in fig. 101, in which the puddle wall is placed in the center of the dam.

The puddle work should increase in thickness downward, 2 inches for every foot of hight over and above the proper thickness at the top water line of the dam. The object of the puddling is only to give security against any imperfection in the rest of the work, through which water might percolate; but if the earth work is properly constructed, of the best materials, it is probable that the water would never penetrate more than a few feet beyond the surface, and would never reach the puddled portion. Nevertheless, puddling should not be omitted, unless under the most favorable circumstances, and even then in no case in which disaster or loss of life might result from a failure, as when a large body of water is impounded in a reservoir. When a sufficient quantity of selected materials has been placed in the dam, the facing on either side may be continued to the proper slope, with any material that will serve the purpose; the inner slope may be finished with soft material, such as peat, or dredging from marshes, when no disturbance by waves, or washing is to be apprehended, but the outer slope should consist of solid matter, which will retain its position, and will not crumble, as for instance, broken stone, rock, or shaly soil.

When an earth dam is thrown across a stream, the most important points to secure are, the waste gates, the outer slope over which surplus water flows, and the foundation. The waste gates should be built in one of the banks, which should be dug away for this purpose, and the frames should be thoroughly well cased in with planking, which should extend some distance into the bank, and be well protected with puddled clay, tightly rammed in around it. The outer slope, and top of the dam, should be of plank or timber, with an apron upon which the overflow is received and carried off. The bed of the river at the foundation should be well searched for sunken logs, or brush, which should be removed before any earth is thrown in. These points should be looked to whether the work be large, or small.

For small streams, dams of very simple construction will be sufficient. Cribs of timber, consisting of a sill, an upright post mortised in the center of the sill, and two timbers placed like the rafters of a house—but one at a greater angle than the other—from the ends of the sill to the top of the post, are placed lengthwise in the bed of the stream for the framework of the dam. The timber which slopes at the greater angle is placed at the front of the dam. The cribs should be placed from 6 to 12 feet apart, as the stream is smaller or larger. They are joined together by planks or timbers, spiked or bolted to them, and the rear of the dam is covered with plank jointed and fitted together very closely. Stone or gravel may then be thrown in, until the cribs are filled, care being taken to pack clay closely at the bottom, so that no water will escape. The front of the dam may then be planked over, and an apron, or floor of plank, laid to protect the bed of the stream from the waste water. The earth-dam upon the banks of the stream must be carefully joined to the crib-dam, and should be supported in the center by posts driven in the ground, to which three or four planks in hight are spiked. A dam of this kind may be made to serve for streams of any size, as it admits of expansion in length, hight, and width, and increase of strength, indefinitely. Where the hight required is not more than four feet, a dam of earth, brush, and logs may be made to answer the purpose. There can be no better principle of construction adopted for such dams, than that made use of instinctively by those sagacious dam-builders, the beavers, whose works, able to withstand floods and freshets, easily made and easily repaired, last for ages, and mock in their simple

Fig. 103.—BRUSH AND LOG DAM.

strength many of our best engineering works. These dams have a foundation of mud and brush, which bind together very intimately; the brush always being laid with the buts down stream, arrests all floating or suspended matter which is brought down with the current, and thus adds daily and constantly to the material, and the strength of the dam. Into this brush is interwoven logs and sticks, limbs and stems of trees, and stones, so placed that the pressure of the water tends to hold them down, and the interstices are filled in with earth which is also

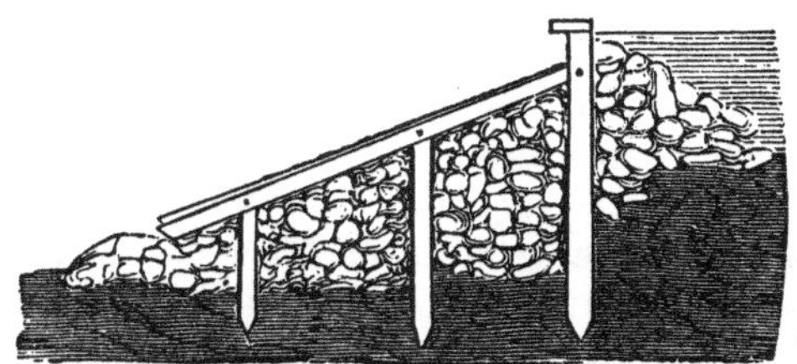

Fig. 104.—DAM OF PILES AND ROCK.

thrown upon the submerged ends of the timbers. For dams of no greater hight than a few feet, and of a length of from 50 to 100 feet, there is none more simple, useful, economical, and permanent than this. For streams the bottoms of which are soft, sandy, mucky, or muddy, this style of dam has no superior. A section of a dam of this kind is shown at fig. 103. A dam made of piles and rock is shown at fig. 104.

This form of dam is suitable for river beds in which piles can be easily driven, such as those consisting of quicksand, mud or soft earth, upon which a structure not founded upon piles would be neither substantial nor permanent. It is made by driving, across the river, three rows of piles, of graduated lengths, as shown in the engraving. These are connected by stringers, solidly bolted to them, and the framework is stiffened, where necessary, by girts and braces. Cap pieces of flattened timber are bolted on to the top, and the whole frame is filled in with rock, and then planked over. The face may be filled in with fine brush, and earth, to a proper slope. Dams of this kind may be made of great lengths, where the fall is not more than 10 or 12 feet, and resist the most severe freshets.

In taking the water from the stream, it is necessary to consult the laws which control the motions of liquids, else counter-currents or eddies may be established, which will wear away, or undermine the dam, or sluice. The dam should slope away at an angle toward the sluice, so that the current of the stream may be easily diverted into the canal without reflux, or regurgitation. To further this end the dam may be placed diagonally across the stream, or partly so, and the floor of the dam should be carried so far up the stream as to cover the entrance to the canal, and a few feet into it, so that the bed may be protected from washing by the current.

PUMPS.—The use of steam pumps in irrigation, will probably be found profitable within a few years, when the valley lands that are easily and cheaply irrigated, are supplied with water. The surplus then running to waste will eventually be raised to the higher lands, by whatever power may be cheapest. In many localities there are no valley lands, but the banks of the streams are abrupt, and if the water is used at all it must be elevated. When it is considered that one bushel of coal contains a latent power within it, sufficient to elevate to a hight of one foot, 50,000,000, (fifty millions), pounds of water, or a less quantity to a proportionately greater hight, the future probabilities of the use of steam pumps in irrigation, will not seem to be misjudged. All that is necessary is to consume the coal beneath a boiler, and apply the power of the steam in the most economical manner, with the best constructed engines and pumps, to the work of bringing the water where it is required. At the present time water is thus procured by one farmer at least, in California, who employs a steam engine and a pump, to raise water from a well, for the irrigation of his crop of vegetables for the Sacramento market. The high price procured for his product, is offset to some extent by the high price of coal, which costs in that locality from $18 to $20 per ton. Where coal is much cheaper, the gain would go to offset the probable lower prices of the product. Yet in many localities, where market crops are raised, it would undoubtedly pay to employ steam power to raise the water, either from streams or wells. If from wells, reservoirs or tanks would be required, both for the purpose of gaining the necessary head for distribution, and for the warming of the water.

There is a large variety of pumps that may be used for this purpose, that are of great capacity.

No mechanical power that we possess is so cheap, or so effective as steam. The effective energy contained in one bushel of coal being able to elevate six million gallons of water one foot high, or a million gallons six feet high, or a hundred thousand gallons 60 feet high, it becomes only a question if the cost of coal and that of the application of the power, will enable us to use it profitably. It is not to be doubted that in some cases now, and in numberless cases in the future, the possibility of the use of steam in agriculture, and especially in irrigation of arable lands may become usefully available. When we see that the consumption of one bushel of coal, costing 20 to 40 cents, in a day, will irrigate 22 acres of land continuously, and as much more than that as the continuity is broken and the consumption per acre is lessened, it becomes very clear that there are many cultivators of the ground that could now make the use of pumps driven by steam to pay them handsomely.

Our present mechanical appliances for raising water are

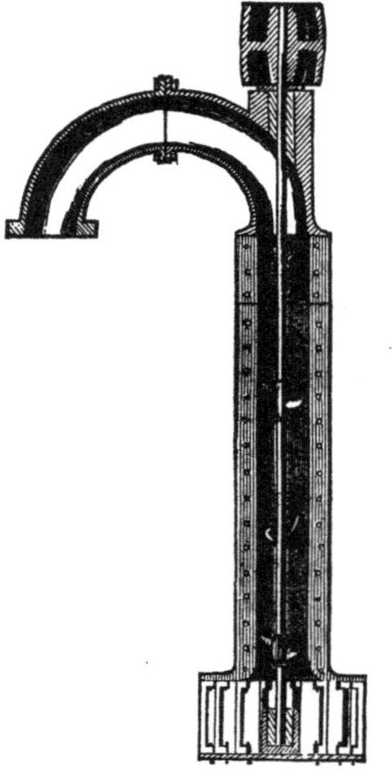

Fig. 105.—CENTRIFUGAL PUMP.

Fig. 106.—METHOD OF OPERATING A CENTRIFUGAL PUMP.

very wonderful. The great rotary pump which discharged the enormous cascade of water at the Centennial, which astonished every visitor to that remarkable display of mechanical powers, is able to throw 100,000 gallons per minute. This would supply about 7,000 acres of land with water for continuous irrigation. The principal upon which this powerful pump works is that of the common propellor of the steam ship. An ordinary propeller shaft is enclosed in an iron pipe, and is rotated by means of a pulley and a belt from an engine. A section of this pump is shown at fig. 105. It is known as Shaw's Compound Propeller Pump, and is manufactured in Philadelphia. The method of its operation is shown at fig. 106. It has the advantage that it can lift water any desired hight by proper adjustment. Perhaps no pump is better adapted to extensive irrigation than this.

It is, however, at the present time, the smaller pumps that will be most available for watering crops at intervals when rain is inadequately supplied. To have then a re-

Fig. 107.—WHITMAN & BURRELL'S STEAM ENGINE AND PUMP.

source that can be drawn upon will be invaluable. For such purposes smaller pumps are made, the cost of which is comparatively trifling. One of these, intended to be operated by steam, and known as the Fairchild Steam Engine and Pump combined, is manufactured by Messrs. Whitman and Burrell, of Little Falls, N. Y., for the very moderate cost of $75. This pump will raise 30 gallons a minute, which will be sufficient to cover 2 acres of land an inch deep every day. The engine is of 2 horse-power, and requires a boiler of equal capacity. The whole, complete, will cost but little over $200, a sum, which considering the inexpensiveness of its operation, is within the profitable employment of almost every market gardener, or fruit grower, who cultivates 10 to 12 acres. This combined pump and engine is shown at fig. 107.

A force pump, designed for house and farm use, but which is usefully applicable for irrigation of gardens, is shown at fig. 108. This is the "Blunt's Universal Force Pump," made by the Nason Manufacturing Co., of Beekman St., New York. A careful examination of this pump, as to its manner of manufacture, and effectiveness in use for the purpose of light irrigation has been entirely sufficient to show its very great value as a cheap and effective pump. Being simple in structure, any person can take it apart, and put it together if necessary; its strength gives it the durability needed for this work, and being furnished with a very capacious sand-strainer, it may be used to pump river or other water which may be muddy, or have sand in suspension, without the least injury to the interior parts. Not the least

Fig. 108.—BLUNT'S FORCE-PUMP.

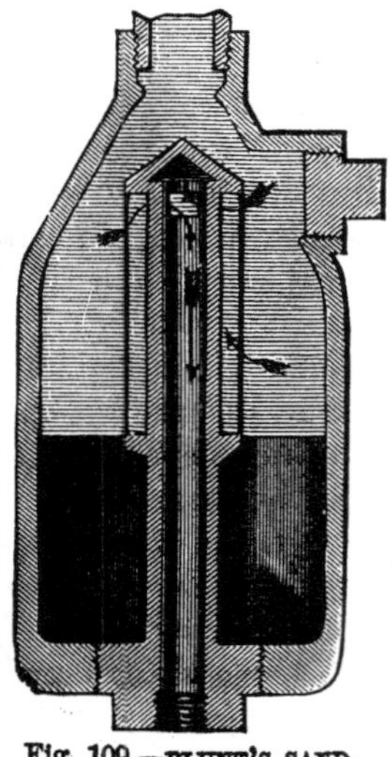

Fig. 109.—BLUNT'S SAND-CHAMBER.

of its value is that it may be put to this use while it fills the place of a house or barn pump, or both. It may be worked by hand, or attached to a windmill or steam-engine. It has attachments for pipes or hose at the spout, or these may be made beneath the surface. The sand-strainer (fig. 109) may be attached to this or any other pump. While there are a great variety of pumps that may be turned to the uses of the irrigator, yet these undoubtedly meet all the requirements of those who may be called upon to use them, from the greatest operator to the least.

RESERVOIRS.—A vast amount of irrigation has been done, and may be done, by the help of storage reservoirs, in which the rainfall of a part of the year is impounded for use during the dry season. The most prominent examples of these storage reservoirs are in India, where ancient works exist, which surpass in immensity, and solidity of construction, what are usually considered as the wonders of the world. The people of India, 100 millions of which depend for their existence upon the water supplied by these reservoirs for the irrigation of their land, have taken advantage of every valley, ravine, or nook, large and small, and have converted them into storage reservoirs, by throwing across them banks of earth, in which the water supply is husbanded, so that none may run to waste. In fourteen districts of the Madras Presidency alone, no less than 43,000 irrigation reservoirs are recorded by the Indian Government as being in effective operation, while at least 10,000 have fallen into disuse. The average length of the embankments is half a mile; one of them, now no longer in use, extending for 30 miles, and enclosing a space of 80 square miles, or over 50,000 acres. The second largest, which is still in use, has an area of 35 square miles, and a dam 12 miles in length. Curious statisticians have calculated that these Indian embankments contain altogether as much earth as would serve to encircle the whole earth with a belt 6 feet in hight and thickness. One embankment of solid masonry, strongly cemented together and covered with earth, exists in Ceylon, which is 15 miles long, 100 feet wide at the base, slopes to a top width of 40 feet, and extends across the foot of a spacious valley. In Europe there are many reservoirs for irrigation. In Spain there are a large number; Italy has most of any European country; in France there are many of considerable extent, one contains 500,000 cubic yards, another 4,000,000 yards, and hundreds contain from 20 up to 50,000 cubic yards. In our own country, where we have seen a railroad system so vastly and successfully extended, it cannot be doubted, that at some time not far in the future, equally costly and valuable works may be constructed, having for their object the reclamation of fertile soils from aridity by bringing to them a supply of water which now flows away uselessly. By impounding the winter rainfall of thousands of valleys, or the melting snow from thousands of hills, floods may be prevented and a store of water be accumulated for use in the rainless season, which may bring into productiveness millions of acres of now waste lands. The manner of making these storage reservoirs is to throw across the outlet of a valley or of a series of valleys connected together, a dam of sufficient hight and strength, furnished with outlet pipes, which discharge, either constantly or intermittently, into a canal. The proper construction of the dam has been already treated of; it will now only be necessary to consider some important and pertinent characteristics of the valleys themselves.

At fig. 110 is shown a system of valleys, which have but one outlet at the narrow neck where the dam is thrown across. This is a typical example of a most favorable opportunity for constructing an irrigation reservoir. In some cases it may be necessary to make more than one dam; some subordinate ones may be required to prevent overflow at lateral points. This will be discovered when the contour lines of the level of the valley are run, as they should be for every three or six feet of

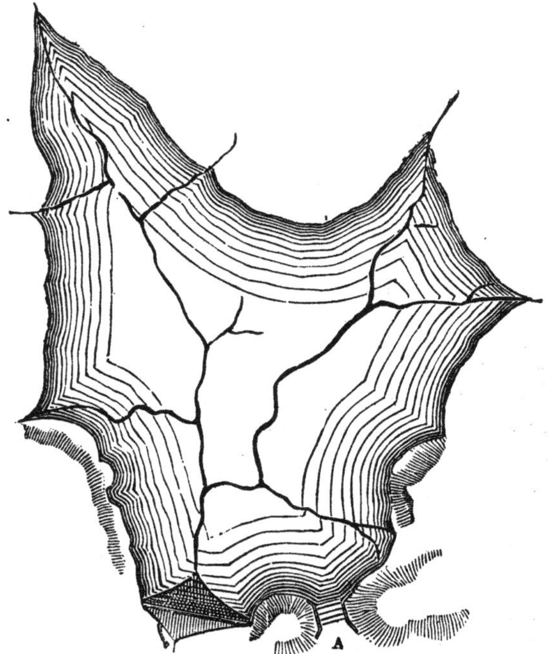

Fig. 110.—A VALLEY RESERVOIR.

elevation. From these contour lines the capacity of the reservoir may be calculated for each level. In the case here illustrated, it will be observed that two valleys run together, and meet where two spurs of high land approach near each other. This combination of favoring circumstances frequently occurs in mountain regions, or among the lateral spurs and foot hills of higher ranges. In the mountain ranges and hills which lie within our territory most subject to aridity, opportunities occur for constructing such reservoirs on the grandest scale, at the most moderate cost. Deep, narrow cañons, which open out into extensive, and sometimes vast valleys, now of little use, for want of soil and on account of their rocky surface, might be easily and cheaply closed, and thus a reservoir of great magnitude might be made. The normal

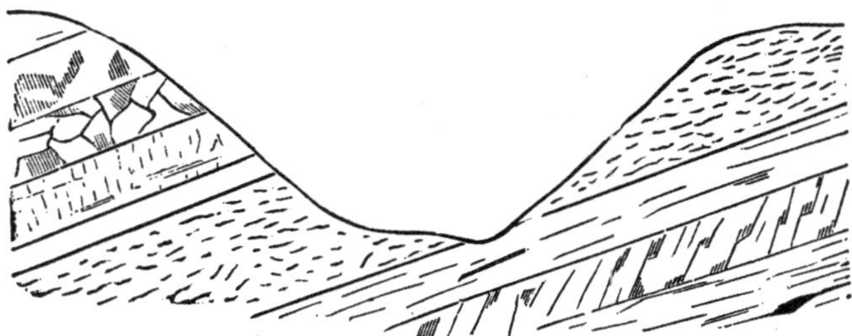

Fig. 111.—VALLEY IN INCLINED STRATA.

flow of the issuing stream might thus be regularly maintained, and destructive torrents from "cloud bursts," and rapidly melting snow banks, might be prevented. But before any expenditure is made in such operations, the geological character of the valley should be examined, lest from unfavorable conditions failure might ensue. This will clearly appear by a glance at the three annexed illustrations. At fig. 111 is shown a section of what is known as a valley of erosion, situated in an inclined formation. It is apparent, that if such a valley be dammed, the water might escape through any one of the strata on the lower side, that might happen to be porous. In this case failure might be expected.

At figure 112 is shown a section of a valley occupying an anticlinal axis. It is equally apparent here, that the escape of the impounded water might be looked for, and that upon both sides if any of these strata be porous. Failure would be certain in this case also.

At fig. 113 is seen a section of a valley occupying a synclinal axis. It is apparent that a reservoir formed in such a valley could not leak by any possibility, even though all

Fig. 112.—AN ANTICLINAL VALLEY.

the strata were porous. In addition to this, a valley of this character will almost always have abundant springs issuing from its flanks, while the previous one can have none at all, and the first mentioned can have them on but

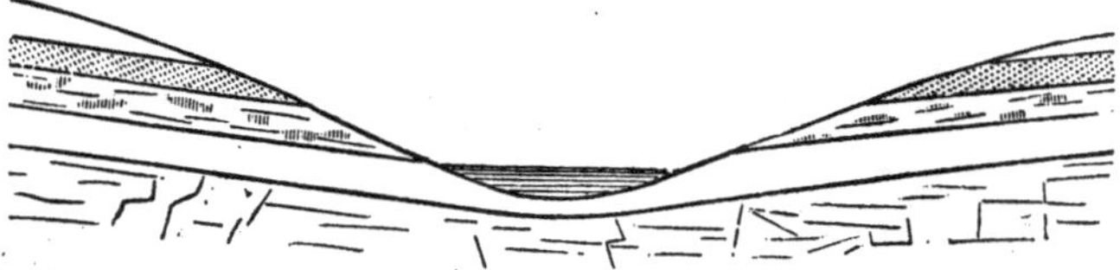

Fig. 113.—A SYNCLINAL VALLEY.

one of its sides, and what may be gained in this way, may be more than lost in another.

The surplus overflow from a reservoir, should be made to discharge at a point away from the dam, as shown at *A*, in fig. 110. This is necessary or at least advisable, as the dam may be damaged by the overflow; or lest to provide the requisite strengthening to resist erosion, the cost may be augmented unnecessarily. A waste-way may be formed in a depression in the edge of the basin, either by excavation, if it is already too high, or by masonry if the existing depression is too low. In case of rupture from any cause, the main work will remain intact. In addition to the waste-sluice, the appendages of a reservoir consist of the apparatus for the discharge of the water, which include the pipes, the valve tower, and the culvert. For convenience and safety, in

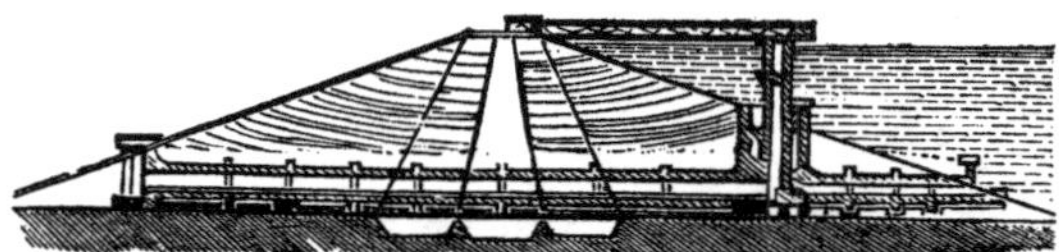

Fig. 114.—DAM WITH CULVERT AND TOWER.

case of the giving way of a joint in the discharge pipe, this should be carried out through a culvert of masonry, of sufficient size to admit a man. This culvert communicates with the valve tower as shown in fig. 114. The valve is a circular plate, which slides between two flanges within the pipe, the surfaces which come into contact being ground to fit accurately together. This is raised by means of a screw attached to a rod having a horizontal wheel for turning it at the top. A form of valve frequently used is shown at fig. 115, the section of pipe containing the valve being bolted by the flanges to the discharge pipe. A valve, in common use in Italian and French irrigating works, is shown in section at fig. 116. This, *A*, may be made of wood, shod at the foot with a plate of cast iron, ground to fit another similar plate attached to the opening of the pipe, *E*. It is raised by the rod, *B*, keyed to the upper part, and is guided by means of eyed wings, *D*, *D*, which work up and down upon the rods, *C*, *C*.

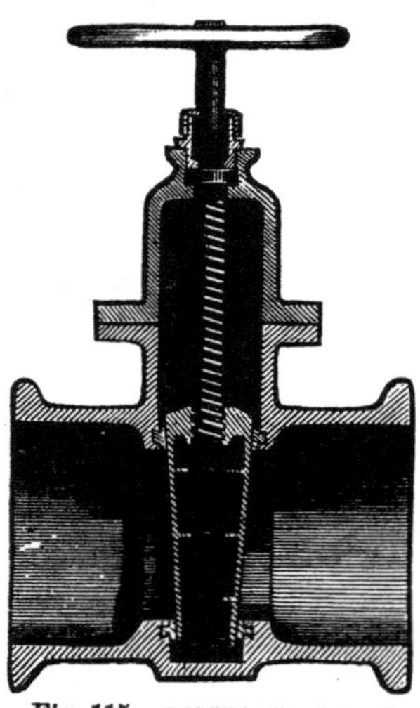

Fig. 115.—DISCHARGING PIPE VALVE.

The hight of the dam above the crest of the waste-weir should differ for different depths of the reservoir. When the dam is 25 feet high, the waste-weir should be 4 feet lower, and for every 25 feet of additional hight of dam, the difference should be increased one foot. The size of the waste-weir should be proportioned to the quantity of overflow to be carried off. This is a matter for calculation of the amount of rainfall and the extent of the area supplying the reservoir. It would always be safe to form a temporary dam of flash boards, or earth,

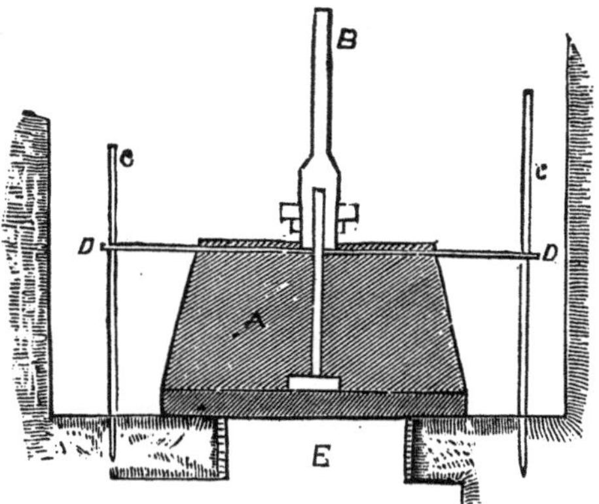

Fig. 116.—DISCHARGING PLUNGER VALVE.

upon the top of the waste-weir, which would raise the surface of the water to the extreme limit of safety, when, if this were overflowed, it would be carried away, and the safe level quickly restored. This is a common practice in India, where a large waste-weir is essentially necessary, on account of the sudden torrents of rain which fall at certain seasons, and where the necessity for saving every gallon of water is paramount. In this way the safety of the works is secured against sudden and unexpected accident. But this possibility of providing for an eventuality, which brings danger with it, in this manner, should not lead to the neglect of carefully gauging the excess of water to be carried off, at times when but little is used in irrigation, and of providing ample accommodation for it.

Reservoirs of smaller size, for use to a limited extent, or for farms and gardens, may be made in a much more modest manner. Where the surface of the ground is

Fig. 117.—RESERVOIR ON LEVEL GROUND.

level, the reservoir may be made by digging out the bottom, and forming the banks of the excavated earth, as shown at fig. 117. A reservoir upon sloping ground may be made by throwing to one side earth excavated from the bottom, and forming the bank, as shown at fig. 118. A reservoir in a natural hollow may be made, by excavating the bottom, and using the earth to raise the sides, as

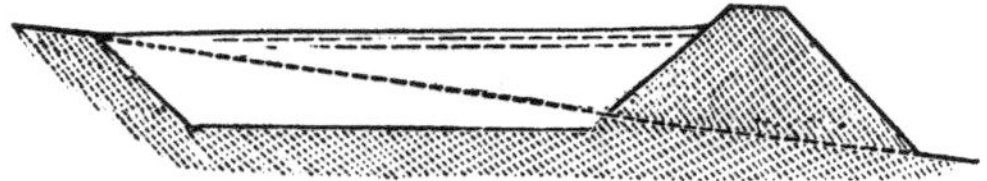

Fig. 118.—RESERVOIR ON SLOPING GROUND.

shown at fig. 119. In these examples, the original outline is shown by the dotted lines, and the finished work by the shaded portion. The scope for the use of such small reservoirs as these, by farmers or gardeners, is in reality very extensive.

It is a question of profit solely. Will it pay for the farmer or gardener to be master of his operations? Will the cost of reservoirs, and of the necessary preparation of the surface of the farm, to make the application of the reserved water possible, overbear the value of the crops raised? With our present defective agriculture, and our consequently unprofitable crops, the necessary cost may in many cases prohibit the improvement. But it cannot always thus remain. The exigencies of a rapidly increasing population will sooner or later compel a different system of agriculture; there must be more enterprise, a greater employment of capital, new methods of producing food, and compelling the soil to yield its maximum crops. One of the improvements will surely take the shape of equalizing the supply of water. There is an extensive scope for profitably doing this now, if we will

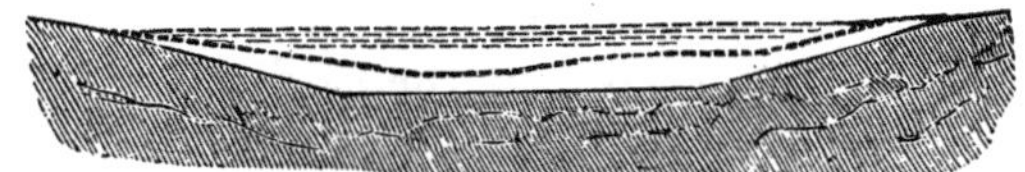

Fig. 119.—RESERVOIR IN A HOLLOW.

only make the most of what opportunities we have. There are numberless farms through which, every Spring, a flood of water pours from the ground upon a higher level. Numberless streams are torrents in Spring and dry gullies in the Summer and Fall. By individual or associated effort, reservoirs more or less capacious, might be made to catch all this useless or injurious water, and make it serve a useful purpose.

ARTESIAN WELLS.—The operation of an artesian well may be explained by the illustration, fig. 120. In this is shown a basin-shaped deposit of various strata, either of rock or of clay, gravel or sand, resting one upon another. One of these strata consists of porous material, lying between two impervious strata; it may be that the one consists of sand or gravel, lying between two beds of clay, or it may be of fissured sandstone, or limestone, placed between two beds of compact rock. At the outer edge of the basin these strata reach the surface. The softer materials being easily worn away, may form valleys through which streams may flow, and a large portion of their contents may escape down the porous bed until the basin is filled. In some cases, when streams are thus situated, the whole body of water sinks out of sight, and flows in an underground channel, until it breaks out in copious springs here and there, or in a body at one place. This happens in well known cases in the limestone regions of Kentucky, West Virginia, Florida, and in Texas, where large streams thus suddenly disappear. In other cases considerable streams or lakes pass over or lie upon such porous strata, and a large quantity of water escapes from them. Let it be supposed that, in the diagram given, a stream or lake is situated at the point *a*, or otherwise that the rainfall of the locality here sinks into the ground and disappears. The water passes through

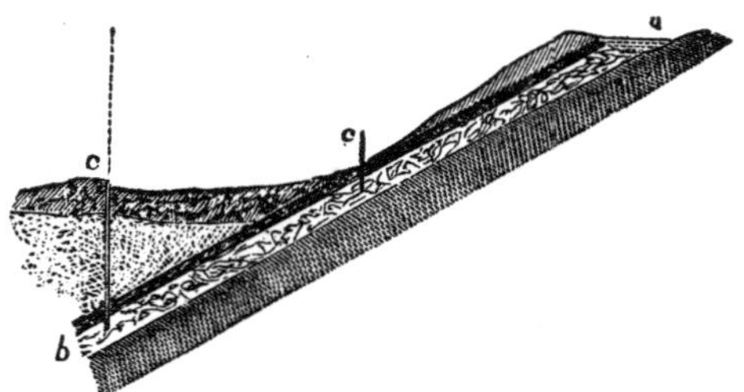

Fig. 120.—PLAN OF ARTESIAN WELLS.

the porous stratum, shown by the pebbled shading, (*b*), until the basin is filled. Then if at any point within the basin, *c*, *c*, a well be bored until the porous water bearing stratum is penetrated, the water is at once forced to the surface by the pressure, and if confined in a pipe, would rise until the level of the source is reached, as shown by the dotted line. This would be an artesian well.

It is evident that a combination of circumstances, rarely existing, must be found to furnish a source of water of this character at all, and that there must be an abundant and permanent supply to furnish wells that can yield copiously and permanently. If there is only an accumulated store of rainfall to draw upon, there is danger that it may soon be exhausted, and afterwards that only a limited supply can be expected. If the source of water is inexhaustible, then only can the wells be made permanent. Thus a few wells in a district, may perhaps yield copiously for a while and then fail, or if the number be added to, the supply of water may be inadequate for all of them, and those at the higher part of the basin will cease to flow the first, and afterwards the remainder will act no longer than the supply holds out. It is certain, therefore, that the risk of expending large sums of money in sinking wells of this kind, will be very great, and that as the number in any locality is increased, the risk of failure increases. Further, the expectation of a permanent supply is seen to be delusive, excepting under a narrow range of circumstances. For these reasons, caution should be exercised in making considerable investments, and founding large hopes upon the basis of irrigating farms by the means of artesian wells. More especially should caution be exercised where an extensive district is to be made dependent upon these wells, and a large number of them are to be sunk in contiguous places.

CHAPTER XIX.

SUPPLY CANALS AND THEIR CONSTRUCTION.

The proper location of the main supply canals of an irrigation system, is a very important consideration. Upon it depend, in a great measure, the cost of the works and their future efficiency. The first cost of a canal will depend, as a matter of course, upon its length, as well as upon its manner of construction. It may be, in some cases, a matter of little moment whether the course of the canal be in a straight line or should curve with the meanderings of the level, or on the other hand, it may be a very important one. No extensive system of irrigation can be built up in a year, or a few years. All the great works in existence have been the growths of lengthened periods, and have been altered, or improved from time to time. But nevertheless, the construction of a system of irrigation works should not be looked upon as a temporary expedient, that may serve a present purpose, and may be changed as need may arise in the future. This would be a short-sighted policy, and one that would be costly in the end. When works of such a useful character as this are completed, many various interests become involved in their stability, and to change their course or character, might, and undoubtedly would, give rise to damage, disputes, and conflicts. The location of the main works should, therefore, be chosen with every regard to future as well as present requirements. As a general rule, the chief consideration should be to select the location with regard to the most copious supply of water, and the largest amount of territory that may be served by it. The actual supply of water should be ascertained with great care and exactness, lest costly works may be constructed, and afterwards found to be inadequately provided with water. There are not wanting cases amongst our new works, in which this unfortunate dilemma seems to be inevitable. The fall in the course of the canal determines at least two things; one, the amount of water which may be passed through it, and the other, the stability of the banks, or the resistance to the wearing action of the current.

The fall should not be more than one foot in a thousand, except there are the best reasons for departing from these limits. This will give a current of 2 feet per second, or about a mile and a half in an hour. Half of this fall, or $2^1|_2$ feet to the mile, may be taken as the standard for average circumstances. This will give a flow of about one mile per hour.

The fall should be regular from beginning to end, else the current will be more rapid in places where the fall is increased, and this will cause the washing of the banks in the steeper parts, and the deposition of the detritus in places where the current slackens. This, in time, will either destroy the canal or render costly repairs necessary. It may be a question whether it is better to follow a long curve around a hollow, or to carry the canal in a flume or aqueduct directly across it. This question may be decided by considering the cost involved in both plans, and the advantages that may be derived, if any, from adopting the more costly of the two. If there is land that may be conveniently irrigated by following the longer course, that would be a point for consideration. But it should be taken into account that a secondary canal can at any time be made to supply dependent territory, and it may not be advisable to carry the main canal to it, for no other purpose than to supply it.

The character of the soil in which the bed is to be made, should be regarded in fixing upon the location. There are some canals in existence which lose 40 per cent of their water by filtration through the subsoil. It is evident that, in such cases, it would have been prudent at least to have been sure that no better location could have been selected.

After this point has been duly settled, the methods of construction need the most careful study. It will be a wise precaution, that may hereafter turn out to be a great economy, to deposit all the excavated earth upon one side only of the canal. If any increase in its size should ever afterwards be determined upon, it can be enlarged at greatly less cost if this precaution is observed.

The same principles which relate to the construction of the larger canals are also involved in that of the secondary and distributary ones. The following remarks will therefore apply to irrigating canals of every description, and to a great extent, to the furrows; excepting those of the most temporary character. It may be that some repetition of previous statements may be made, but as they will be found pertinent to the matter in hand, no apology may be needed for that.

In the construction of canals of whatever description, so long as their bed and banks are of earth, the inclination of the bed, the size of the channel, its form, and the nature of the soil through which it is carried, are of the utmost importance; because upon these depend its capacity for delivering water; its cost of construction; its permanence in use; and the prevention of loss of water by filtration through its banks or bed. Upon the inclination of the bed depends the velocity of the current and the stability of the banks. It is necessary to limit the velocity of the stream in the canal, lest the banks should be degraded, and washed down into the bed. Water flowing at the rate of 120 feet per minute, which is the rate of flow with a fall of one foot in a thousand, is considered the limit of safety in the most consistent soils. Water flowing at half this speed will wash the banks of a canal made in sand and fine gravel, but it must not be forgotten that the velocity of a stream is greatest in the middle of the surface, and least at the bottom and sides where it comes in contact with the earth. Thus the flow in the center of a wide stream being at the rate of ten feet in a given time, will be only eight feet in a deep canal, and six feet in a shallow one.

The following table gives the different and the mean velocity of streams:

Velocity in inches per second at the surface.	*Velocity in inches per second at the bottom.*	*Mean velocity.*
4	1.	2.5
8	3.3	5.6
12	6.	9.0
16	9.	12.6
20	12.	16.0
24	15.	19.5
28	18.4	23.2
32	16.	26.8

The slope of some of the largest irrigating canals in Europe is from 13 to 200 feet in 10,000. The slope in the canals of the Tyrol and other localities in the Alps is frequently as great as six feet to the thousand. In these cases the sides of the canals are of masonry or timber. The rule upon which the average fall of canals is indicated is as follows, viz.: For a bed which consists of fine mud, 16 feet in 100,000; for soft clay, 45 in 100,000; for sand, 136 in 100,000; for gravel, 433 in 100,000, and for solid clay, 570 in 100,000. With greater inclinations than these there is a probability that the substances of which the bed is formed will be taken in suspension and transported by the water.

Fig. 121.—A DEEP CANAL.

It is obvious that in those cases in which the fall is at the minimum, the size of the canal must be enlarged proportionately to pass a required amount of water. The velocity may be hastened without enlarging the size in certain cases. For instance, it is a rule in hydraulic engineering, that the velocity is in proportion to the mean radius or diameter of the canal, other things being equal. Thus the water in the canal deep in proportion to its width, as illustrated in fig. 121, meets with less resistance from the surface of the bed and sides, (called by engineers the "wet perimeter") than that in the shallow canal seen in fig. 122, and its velocity being therefore greater than in one of a contrary character, a larger quantity of water is passed through it in a given time.

Fig. 122.—A SHALLOW CANAL.

In soils that do not admit of rapid currents, and in cases where a greater fall is unavoidable, it is customary to construct the canal in sections, joined by chutes of stonework or timber. The water passes through these chutes with great velocity, and accomplishes the fall without injury to the canal.

The inclination of the banks depends upon the consistency of the soil. The angles of repose, or the slopes at which various kinds of soil will cease to slide down a declivity, are as follows: Wet clay, 16 degrees from the horizontal; dry clay, 45 degrees; coarse gravel, 40 degrees; compact earth, 50 degrees; arable loam or mucky earth, 28 degrees; wet sand, 22 degrees; dry sand, 38 degrees; fine gravel, 40 degrees. It depends upon the position of the canal as regards the surface as well as

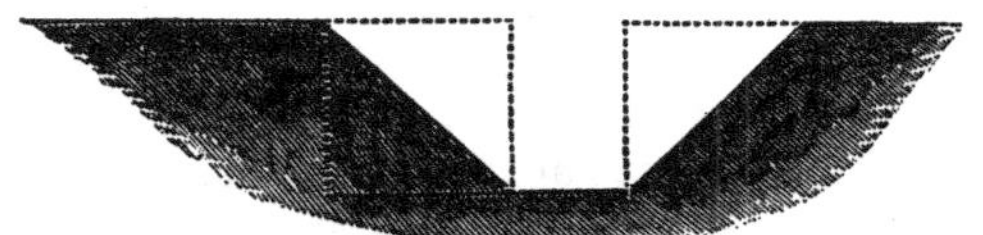

Fig. 123.—SLOPE OF "ONE FOOT IN ONE."

upon the nature of the soil, what inclination is necessary to be given to the banks. When the canal is excavated wholly beneath the surface, an angle of 45 degrees, or a slope having one foot of vertical hight to one of horizontal base, is generally chosen. This is shown in fig. 123, in which the dotted lines show that the slope is the diagonal diameter of a perfect square, and therefore one of 45 degrees, or of "one foot in one." When the canal is deeply excavated, the slope should be broken by a narrower bank, slightly above the level of the water, shown in fig. 124, and the upper slope above the bank should be increased. The width of the bank should be proportioned to the hight of the upper slope; its purpose is to prevent earth loosened from the slope falling into the canal. In carrying a canal around the spur of a hill, the earth excavated should be made to increase the width of the bank, as shown in fig. 125, in which the dotted line marks the excavation and the removed earth. The moved earth will adhere more closely to the old surface

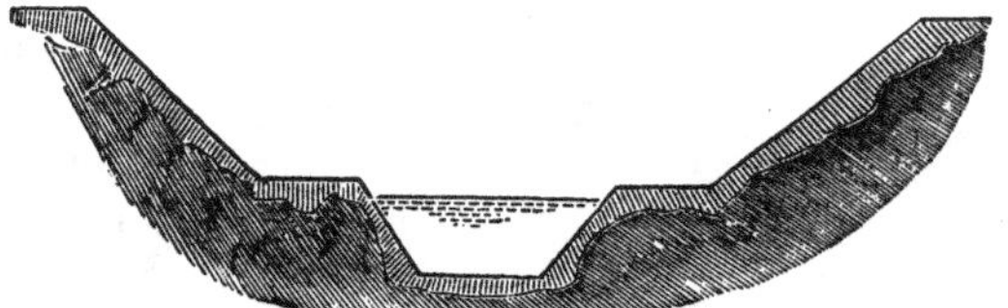

Fig. 124.—FORM OF A DEEP CANAL.

if that is loosened with the pick, so as to secure an intermixture of the new earth with the old. Where the hill-side is of loose earth, it may be necessary to protect it by stonework, laid up as seen in fig. 126. Where a curve is made upon a hillside, it must not be forgotten that the

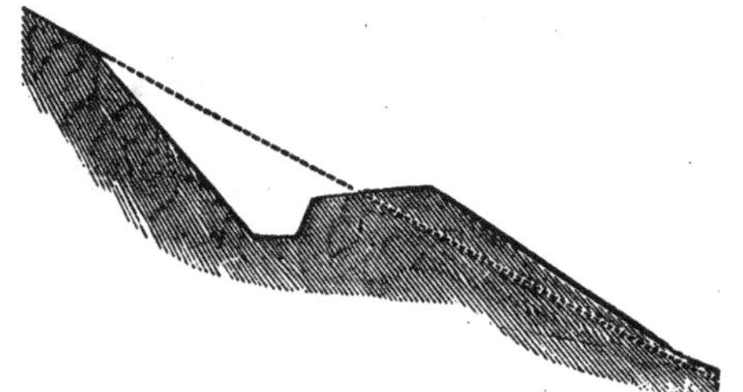

Fig. 125.—CANAL AROUND A HILL.

stream impinges upon the bank at every point of the outward curve. Sometimes, therefore, the stone-work will need to be laid up upon the outer side, as the proper place for it is where the water will wear the most. Protection of some kind will be needed at those points. This may be given by driving stakes and wattling brush among the stakes, with the buts pointing down stream, or by increasing the slope of the banks on the outside curve. Sometimes a canal needs to be carried underground, beneath roads or buildings. A wooden bridge may be made as shown in fig. 127, which, when covered with earth, appears as at fig. 128. It consists of a piece of round timber, to which short planks are strongly nailed by one of their ends. The other ends are spread apart as far as may be necessary to give sufficient capacity to the canal. The bridge is put together in the canal, and when it is finished is covered with earth. Its triangular form gives it great strength.

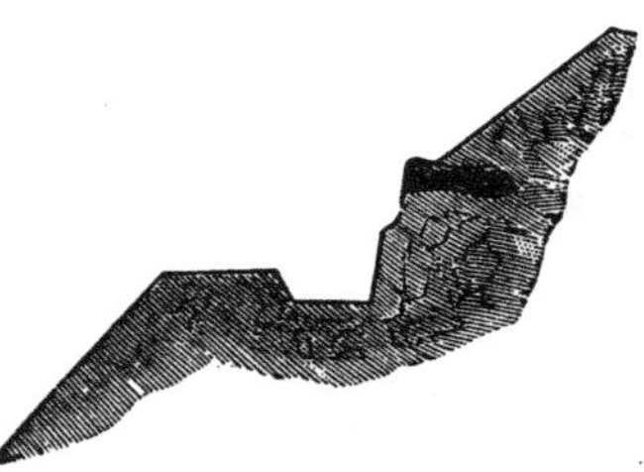

Fig. 126.—PROTECTED CANAL.

Where the soil is porous and open in its character, considerable loss of water often occurs by percolation. This is to be prevented by puddling the bottom and sides with clay or compact earth. The clay is deposited upon the banks, and as it is softened and reduced to a plastic condition by the action of the water, it is carried down and deposited in a layer upon the bottom of the canal. Gravelly and loamy soils may be made water-tight by puddling the moist bottom. This may be done by driving a flock of sheep up and down the canal when the bottom is wet, or drawing logs up and down by horses. When a canal must either cross a valley or be carried around it, it will often be

Fig. 127.—BRIDGE FOR CULVERT.

Fig. 128.—COMPLETED CULVERT.

found profitable to extend its length around the curve. It will be so in cases where the soil of the valley can be brought under irrigation, and where it is compact and capable of retaining the water. Where these circumstances are not present, it will be best to carry the water across the depression by a wooden channel, supported upon timbers, or by an inverted siphon resting upon the surface and covered with a bank of earth, or buried wholly beneath the surface. If an inverted siphon is used, it must be remembered that the confined water exerts considerable pressure, which must be provided for by securely strengthening the tube.

The capacity of the canal is an element which enters, in an important degree, into the calculation as to its construction. To estimate the capacity of a stream of water it is necessary to find the area of the cross section of the stream in feet, and to multiply this by the velocity in feet per second or minute. This should be reduced one-

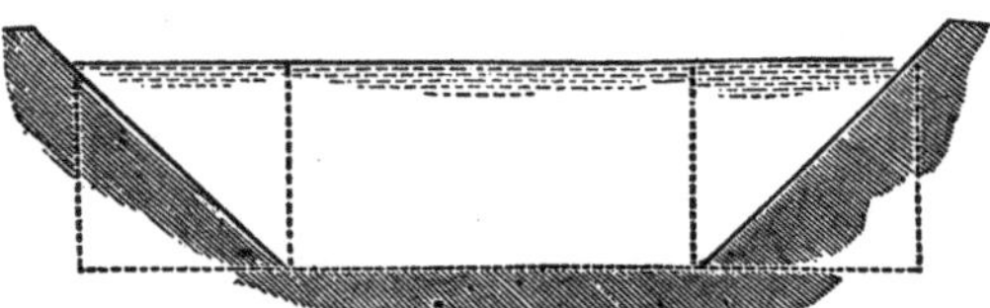

Fig. 129.—PLAN OF MEASURING A CROSS-SECTION.

fifth, to allow for the lesser velocity at the bottom and sides, before explained. The result is the cubic feet of water passing down the canal or river in the time indicated. A cubic foot of water weighs 62$^1|_2$ pounds, and measures about 7$^1|_2$ gallons. To find the cross section of a stream, the figure formed by the surface of the bed and that of the stream is taken and averaged, or reduced to determined geometrical outline. Thus a stream one foot deep in the center, four feet wide on the surface, two feet at the bottom, with banks sloping at an angle of 45 degrees, will have a cross section of three feet. This result is obtained by adding the width of the surface to that of the bottom, dividing by two, and multiplying the sum by the depth. This is explained by fig. 129, in which it is seen that the two triangular side sections of the area of the stream are equal to half the central section. If the bottom of the stream came to a point and had no width, then the two halves would be equal to a square with a diameter equal to half the width of the surface. The velocity is found by floating a cork or piece of light wood upon the stream, and accurately measuring the distance traveled in a minute. The usual estimate of the water required in extensive, continuous irrigation, is one cubic foot or 7½ gallons per second for 100 acres. Other estimates double the quantity of water required, but it is found, as the soil becomes saturated, that much less moisture is required to supply the crops. This smaller estimate may thus safely be taken as the basis of calculation for the size of the main canals.

In calculating the capacity of the canal, or rather the amount of water that will be carried through it, allowance must be made for the loss by filtration, and also by evaporation. The total loss from these sometimes amounts to 50 per cent of the water entering the canal, during a flow of a few miles. The Platte Water Canal Company, of Denver, loses 700 inches of water out of a total inflow at the head of their canal of 1,700 inches. Both filtration and evaporation may be reduced to a minimum by giving to the canal the form shown in fig. 121, by which the bed is narrowed, and the surface exposed to the atmosphere is decreased.

The measurement of the water supplied should be accurate. Generally this is done by means of a gate of given dimensions, fixed in a sluice-way accurately constructed, which is graduated so that it may be raised to a mark designating the quantity of water passing through the opening. The quantity issuing is regulated by the head or hight of the water above the opening, as upon this depends the velocity of the stream escaping. The exact velocity of the stream issuing under a certain head is not ascertained by any arbitrary rule, but is estimated and agreed upon by irrigators as a matter of custom and convenience.

The method of measurement common in this country, is by an opening of so many square inches with a head of three inches of water in the flume. The actual quantity of water which may flow through this opening depends upon many varying circumstances, such as the size of the canal, the substance of which the flume is made, the shape of the flume, its position with regard to the course of the main current, with other modifying influences, all of which may cause differences in the quantities delivered through openings of the same size. By and by, when our circumstances require it, however, some more precise method to arrive at the exact quantity of water escaping through the orifice will undoubtedly be discovered.

The gates for the passage of water into the smaller canals should be carefully made. Wooden sluices are destructible, and can scarcely be made close. If timber sills, sideposts, and plank are used, they should be made of the best of oak. A cast-iron plate should be laid in the sill for the gate to close upon, and the gate should be shod with a cast-iron shoe, beveled and planed to fit the planed surface of the plate. There will then be no leakage. A well constructed sluice-way should have a cast-

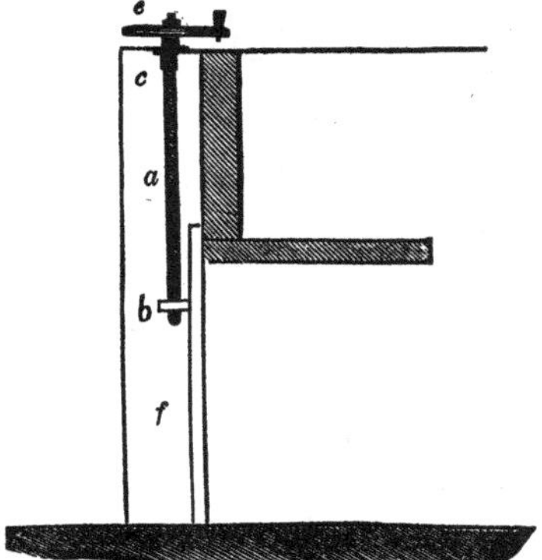

Fig. 130.—SECTION OF GATE.

iron plate in the sill and a cast-iron shoe upon the gate. A common mode of construction is as follows: The gate slides in side quoins; a flat bar supports an upright frame, in which a toothed-wheel is fitted, gearing into the rack of the stem of the gate. By turning the toothed-wheel

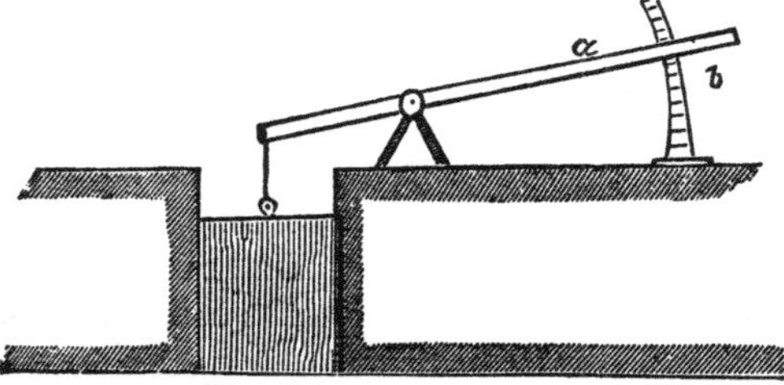

Fig. 131.—A LEVER-GATE.

by a crank, the racked stem is raised or lowered, and the gate with it. At fig. 130, a form of sluice worked by a screw, *a*, passing through a projecting eye, *b*, in the gate, *f*, and a revolving nut, *c*, and lever-wheel, *e*, in the frame. By turning the lever-wheel the gate is raised. Another form is shown at fig. 131. The gate is lifted by an arm *a*, which works on a pivot, and catches into a rack on a quadrant *b*, and is there held, keeping the gate open at any desired hight. A very common wooden stop is shown at fig. 132. This is suitable for small channels, and is made of two boards joined together, but separated at the ends, as shown at *a, a.* In the space between the boards a sluice board, *b*, is passed, being lifted by a hand-hole, and kept at any point by a wedge, *c.* An aperture is made in the boards of a size required, the lower edge being level with the bottom of the channel.

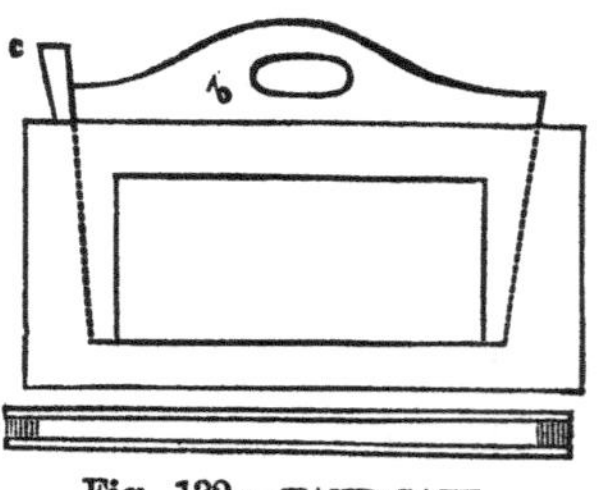

Fig. 132.—HAND-GATE.

Sometimes it is necessary to carry one channel across another at the same level, and yet keep them separate. This is done by constructing a pipe of plank, see fig. 133, in the shape of an inverted siphon, and carrying it beneath the channel to be crossed. This pipe is sunk in the channel of which it is a continuation. At the entrance to the pipe a well or basin, *c*, is sunk in the channel, in which matter suspended in the water may fall and be caught so that it may not obstruct the pipe. In the same manner a canal may be carried beneath a road.

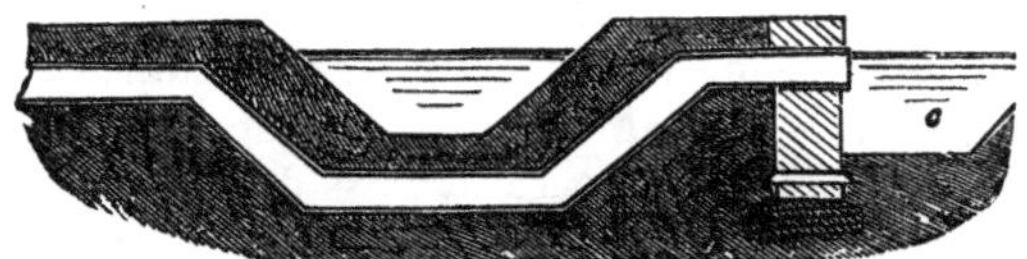

Fig. 133.—MANNER OF CROSSING A CANAL.

In forming the smaller permanent channels, some labor may be saved by taking advantage of natural depressions in the ground, and forming the channels there, using the excavated earth for filling up the depression. Thus in the case illustrated by fig. 134, the earth is removed from the center of the hollow, and placed on each side, raising the sides, as shown by the dotted lines, and leaving the canal of the shape indicated. A case in which a canal is needed upon sloping ground is shown at fig. 135. The earth excavated is placed as before on each side, and the level raised as shown by the dotted lines.

Fig. 134.—CANAL IN A HOLLOW.

The distributing furrows upon cultivated grounds cannot be permanent. They are destroyed at every plowing, and must be re-made for every crop. For the majority of crops that are planted in hills or drills, the furrows between the rows will serve for the irrigation of the crop. For other crops that are sown broadcast, the furrows may be made by rollers, figs. 96 and 97, which press the ground into regular corrugations as soon as the seed is sown, and harrowed.

It may be well to notice in this place the exaggerated and erroneous ideas of some writers upon the subject of irrigation, who do much injury by misleading public opinion upon some vital points. For instance in a paper upon this subject, published in the Report of the Department of Agriculture for 1871, it is declared, "that with very few exceptions, every foot of land lying in Colorado and Kansas, between the base of the Rocky Mountains and Kansas City on the Missouri river, is susceptible of irrigation." It is true that it is *susceptible* of irrigation if the necessary water can be found by this too sanguine and mis-informed writer. But the water is not there, and in fact but a very small portion of the territory can ever be brought under irrigation with the existing supply. As an example, the Arkansas valley may be considered. The nearest crest of the water-shed, on the north side of this river, is 45 miles distant; on the south it is much more distant. If we calculate a territory only 90 miles wide, and 500 miles long, depending upon this river, it would contain 28,800,000 acres, requiring at one cubic foot per second per 100 acres, 288,000 cubic feet per second. To supply this there would be required a river nearly 2 miles wide and 10 feet deep, flowing 2 miles an hour, with no allowance for loss by evaporation and percolation. The Arkansas is not one-fifth of this capacity, and could not supply one-tenth of the territory. What then becomes of the rest of the territory which is without

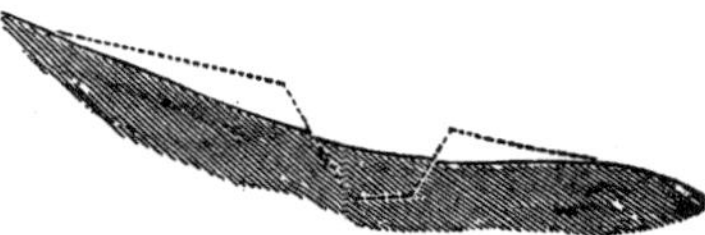

Fig. 135.—CANAL ON A HILL-SIDE.

any great river to supply it? The absurdity of the above assertion is manifest.

Again, this seems a proper place to refer to the erroneous figuring, noticed on page 172, in regard to the canals in California. A canal 55 feet wide, upon the surface, (and averaging only 50 feet in width), 4 feet deep, and flowing 2 miles an hour, can only supply 600 cubic feet per second, or enough to give one cubic foot per 100 acres to 60,000 acres, or half a cubic foot, per 100 acres, to 120,000 acres; instead of supplying 325,000 acres—the capacity claimed for it. But 25 per cent of the supply may easily be lost on the route, and this would serve to still further reduce the number of acres served. It is very important that these calculations should be made with exactness, or some very costly mistakes may be made, which may very reasonably tend to disgust persons who are not well informed, as to the actual cost and merits of irrigation.

The manner of construction of lesser canals, for distributing the water, should be consistent in all respects with the conditions and requirements here pointed out; there is probably no necessity to enter into details which might be tedious, and would necessarily be a repetition, to some extent, of what has been described heretofore.

One point, however, must not be omitted; that is the connections of the earth works with flumes, gate frames, sluices and boxes for the proper conducting of the water into devious courses. It is a known property of water that it is very much inclined to "creep" along the surface of any pipe, arch, culvert or sluice, whether of iron, brick, masonry, or wood, which is imbedded in earth work. The connections should, therefore, be made with great care. As a rule, the walls of gateways, or sluices, should be protected with flanking walls of the same material of which the main works are constructed, or else should be protected by piles driven firmly into the bed and braced or anchored into the bank with timber stays ; and the water surface of the piling should be planked. For small works, piles and wattling of brush may serve a very good purpose to prevent erosion and undermining. But in whatever way it may be done, some protection against the wearing effect of currents or eddies, or the penetration of water into the work should be provided, wherever the course of the water is changed, and a stream is divided or diverted.

For smaller ditches or canals, such as those of six feet in width or less, a grade of one foot to the thousand will be found hardly sufficient. Two feet to the thousand would not be an unsafe inclination for such channels, and where the soil is firm or tenacious, an inclination of three feet might be allowed. The narrower the canal the greater ratio of inclination would be needed.

Caution should be exercised to frequently observe the condition of the banks of secondary canals, when the soil of which they are made is not of a very consistent character, and where the water is confined within embankments. The cutting of a bank of earth by a current of water, is a work which grows rapidly from small beginnings to great proportions, and a break in a bank may be the work of a very short time, if a little wasting is allowed to pass unchecked. The damage that may easily be done in one short hour, by the escape of the water of a ditch carrying but a square foot and a half, would easily surprise one unused to such effects, and might be irreparable for a whole season. While the irrigator is greatly benefited by the water he uses, so long as he can control it in his service, he is always liable to be damaged if he permits his servant to escape control and become his master. This, however, can only happen by inexcusable negligence or mistakes arising from inexperience.

CHAPTER XX.

THE APPLICATION OF WATER TO THE IMPROVEMENT OF LANDS.—SILTING OR FERTILIZATION OF LANDS BY FLOODING.—RECLAMATION OF SALT MARSHES, RIVER FLATS, AND SUBMERGED LANDS

The methods of irrigation described in the preceding chapters, have for their object the supply to the soil of water sufficient for the growth of various crops, either during the season of heat and drouth, or in climates in which the rainfall is not sufficient for the needs of vegetation. But there are methods of employing water in the improvement, reclamation, or indeed, the actual making of land, that belong to the art and practice of irrigation which claim at least some notice in this work. It is very probable that many years may elapse before the gradual growth in value of our agricultural lands shall arrive at that point, which will make it desirable to make extensive use of these methods of improvement. But there are many cases occurring, in which the owners of lands that are amenable to improvements of the character here referred to, are either putting these improvements into operation, or are anxiously seeking for practicable plans for reclaiming their property. It would be well, however, to caution the owners of waste lands against unwisely undertaking large expenditures of money, before they have consulted some competent engineer who is practiced in this special business, or until they have felt their way by completing some portion of the work in a satisfactory manner.

"*Silting,*" *or fertilizing by flooding with water having much earthy matter in suspension,* is the first of these indirect methods of irrigation to be treated of. This practice depends for its effects upon the presence of much suspended matter in the water used; a large supply of water; a soil that is destitute of fertility in its present condition, either naturally, or as the result of damage by washing or flooding, and that is so situated that it can be covered with water from a muddy stream, from which the load of suspended matter may be deposited during a period of rest. After this has been done, the clear water is withdrawn slowly, so that the newly deposited soil is not disturbed, and a new supply is let on.

The lands that may be thus improved, are obviously only those in river bottoms, or in bends of streams, where damage has been inflicted by the washing of freshets, and upon which water from the stream may be flowed either by damming, or by the high water of floods, and upon which the water may be retained by a series of banks until it has served its purpose, when it may be withdrawn through flood-gates or spouts in the banks.

After the surface has been brought to a level, or to a smooth, regular and not excessive slope, in one direction, the arrangements for retaining the water should be made. A succession of banks, as described in Chapter XII, pp. 126–127, will be needed. The higher the banks, and the deeper the sheet of water that can be retained, the better; for the more water that can be impounded, the greater the burden of soil that will be deposited. The discharge of water must be carefully regulated, lest the deposit be stirred up by the current, and carried off by the retreating waters. To obviate this danger, the gates should open at the top, and not at the bottom. The best ar-

Fig. 136.—WASTE GATE.

rangement consists of a flume of plank, built in the embankment, as shown at fig. 136, in which three or four or more narrow planks are made to fit in grooves. When the water has become clear, and is ready to be withdrawn, the top plank is raised at one end, or is removed altogether, and the water allowed to escape. When the water has reached the top of the second plank, that is removed, and so on until the ground is cleared.

As soon as a deposit has been made, sufficient to bear a growth of grass, the seed may be sown and the operation suspended. It may be repeated again when the herbage has taken root, in which case the management will be precisely the same as that of a water meadow, described in Chapter XII, and the same rules that are there given will be proper for its treatment.

The reclamation of Salt Marshes is a work of draining, primarily ; and would be out of place here, except that the following work, the freshening or desalation of the soil, which is a process of irrigation, is so closely connected with it that the one becomes a part of the other, and can only be carried on in conjunction with it. The importance of the reclamation of the millions of acres of salt marshes along the coasts, is so highly considered by thoughtful persons, that at a recent meeting of a scientific society at Boston, this was stated to be one of the chief means of the recovery of the agriculture of Massachusetts to its former vigor and profitable success.

The drainage of salt marshes consists in embanking them from the tidal flow, in draining the waters from the marsh into ditches, from which the escape is by means of sluices with gates which permit the outflowing water to pass, but which close themselves against a flow from without. A gate of this character is shown at figure 59. As soon as the salt water has been diverted from the land, the work is but begun ; for the soil, saturated with salt, produces no herbage but coarse sedges, reeds, or other sea-side plants. Generally there is an abundance of fresh water available for the improvement of the marsh, but in the effort to keep the salt water out this is kept in, with the result of perpetuating the marsh, notwithstanding its drainage. The remedy is by flooding the land systematically and as copiously as possible with this fresh water, and then withdrawing it; repeating the process until the salt has been dissolved and carried off.

The remedy can be applied in two ways, at least. The one is, in case a stream of water passes through or by the marsh, when the fresh water is diverted by a dam in the stream and a canal or ditch, upon the salt land, where it is retained for a time and then discharged through the gate at low tide. Another is, by closing the gates and securing them so as to retain all the drainage water until the ground is deeply covered, when the gates are opened and the water discharged. The repetition of this process will, in time, remove the salt from the soil and leave it ready for the plow, and the profitable cultivation of crops. To carry out these operations effectively, it is only necessary to apply to practice any of the methods, found to be most advisable, that are explained and described in the preceding chapters of this book. It is difficult to imagine a case, in which it would be impossible to apply some of the plans herein described for the drainage and irrigation of meadows.

The Improvement of River Flats, that are partially or periodically submerged, is another of the direct operations of irrigation. The object of this improvement is, to reclaim low lying banks of gravel, sand, or mud, either upon the sides of tidal estuaries, or upon streams that have changed their course, and have left these ruined spots to mark the ravages made by former freshets. This process of reclamation consists in forming banks or courses of piles and brush, by which the tidal flow or the high water of rivers at certain seasons, when a large quantity of suspended matter is carried down, is arrested or retarded, and made to deposit its burden.

When land is to be thus reclaimed, the first thing to be looked to is the nature of the outfall for future drainage, when the newly made ground requires to be dried and made fit for cultivation ; the second is, to be sure that the amount of solid matter carried in suspension by the stream, is sufficient to warrant the expectation that the process will be completed in a reasonable period of time, and at a cost that will not surpass the probable future value of the land. When these points are decided favorably, the next thing is, to choose the method by which the work may be done ; as one method may be used, by which eight or twelve years may be required to do the work which may possibly be done in two or three years, by another method. Thus, by simply retarding the flow by cross-lines of stakes, with brush wattled between them, or by coarse basket work or gabions anchored with stone and deposited in lines, which is the least expensive plan, some years may elapse before the ground may reach the hight of ordinary high water, and become solid enough to sustain an embankment ; when by throwing up banks of mud upon foundations of piles and gabions filled with earth or gravel, and making sluices so as to enclose the muddy water and retain it until its load has been dropped, when the clear water could flow off, a depth of soil of from one foot up to four or five feet has been gained in one year. Generally the process is a very slow one, and before the work is undertaken some trustworthy estimates should be procured, as to the cost of the work, and the probable length of time that may be required for its completion.

The erratic course of rivers and their fickle behavior when in flood, is an element that deserves close study. Much of this depends upon the geological character of the banks, as well as upon the velocity of the stream.

For instance, it is a well established fact that, while coarse gravel resists a current of two miles per hour, fine gravel is moved by a current of one mile per hour; ordinary sandy soil by a current of half a mile per hour, and fine mud is carried away by a current that is almost imperceptible; so that the abrading action of flowing water depends upon both of these contingencies. When a stream, flowing with sufficient velocity, meets with a soft spot in the bank, it soon excavates a concave outline forming a bend, around which the current sweeps and is deflected with violence against the opposite bank. Cutting away then begins in a new place, and a second bend is formed here. The effect is continued, the banks are hollowed out in opposite directions, the river, deflected from bank to bank enlarges the bends and lengthens its course. But as the course is lengthened, the fall is reduced, the velocity is decreased, and the destructive stream becomes a placid, gentle, harmless current; incapable of inflicting further injury upon its banks. Besides, in time of floods, the broad stretches between the bends are swept over by the spreading stream, and the wide course permits the waters to escape with rapidity, and without dangerous velocity. When, therefore, it is determined to reclaim one of these broad stretches, over which the water flows, it must be remembered that it is an effort to return to the former conditions when the river was an active and destructive agent. The work, therefore, requires to be done with care, caution, and skill; lest a new course of destructive action be caused,

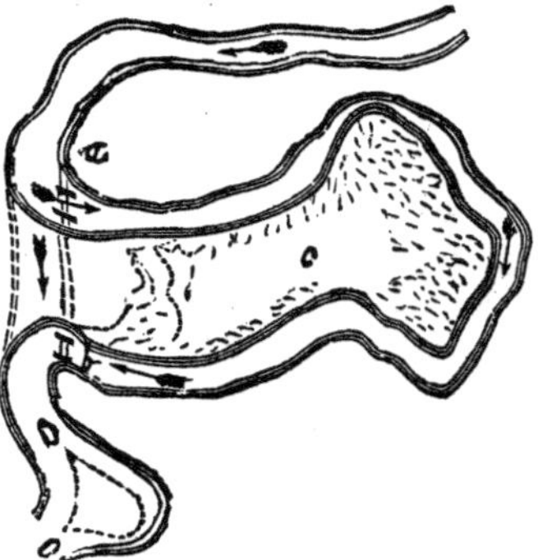

Fig. 137.—PLAN OF A CUT-OFF.

which may seriously injure the banks below, and totally upset the slowly acquired stability of the stream. An illustration of the channel of a river, that has established a winding course, and has formed bends or flats that may be brought under improvement, is given in fig. 137. This

Fig. 138.—SECTION OF CUT-OFF.

may be either a tidal river, or otherwise. The course of reclamation of the extensive tongue of land, surrounded by the bend, will be the same in either case. Here is an excellent opportunity for a cut across the narrow neck, as shown. The cross section of the cut, with its embankments, is shown at fig. 138. By this cut the current is diverted from the bend, which at times of flood may be covered with muddy water, and gradually silted up. Two gates are made in the right bank of the stream, as shown

Fig. 139.—MODE OF PROTECTING THE BANK.

at *a*, *b*, fig. 137. The in-flow gate is at *a*, and the out-flow at *b*.

The banks of the cut should be protected from the abrading action of the increased velocity due to the greater fall, by means of rubble stone, retained in place by piling and planking, or by the piling alone. Bundles of brush may be used in place of stone, and covered with earth, as shown at fig. 139. Nothing tends more to the permanence of a river's banks, than a smooth surface upon which the water can find no irregularities to beat against, but from which it glides gently. In forming the protecting banks, it is best to place them at such a distance from the bed of the river, as to leave a solid fore-shore. No angles or bends should be made, but the lines should either be straight, or in easy curves. The materials should be such as will bind firmly together; a mixture of clay and sand being the best. Combinations of masonry and earth-work should never be used, as no proper bond or union can be formed between them.

The surfaces of the banks should be covered with grass as quickly as possible, and no trees should be planted upon artificial embankments. If water passes beneath an embankment, a trench should be sunk, and filled with clay puddle. When one side of a river is protected, the other side is greatly endangered, unless equally guarded, and the protecting works should therefore be made upon each side.

Where the formation of a new cut is not possible, as upon banks in tidal rivers, or in streams only one bank of which is owned, or in estuaries, the method of staking or piling, should be adopted. This consists in driving piles or stakes, in double or single rows, across the tract to be reclaimed; or in dividing it into sections by cross lines of stakes or piles with brush interwoven, or by making deposits of stone along the lines of stakes. By these methods the current is retarded, eddies are formed, and the water is rendered stagnant; in either case, any suspended matter is dropped within the lines of the obstructions. As the surface rises, additional stakes are driven and more brush is placed between them, and weighted down with stones, until the level is raised sufficiently to warrant the exclusion of the water by more solid structures. Runs or water-ways, by which the receding water escapes, whether it be the tidal flow or the water of rivers, are to be filled by running lines of stakes across them, and filling between them with brush or stone. These cross lines should be made lower in the center than at the ends, lest the water should escape around either or both ends, and form new channels. When deep gullies have been formed, a different course must be pursued, viz.: to throw into the deepest part coarse basket work, gabions, or bundles of brush, which are loaded with stones, until the bottom is gradually raised; when it will become possible to use the stakes and brush. When the level of the made ground reaches the usual high water mark, it is ready to be enclosed between banks, and rarely before this point is reached. The course to be followed is then such as has been already described in this chapter.

The surface having appeared above water, and having been embanked and freshened, as previously described, it is prepared for cultivation by being sown to grass as a preliminary proceeding; for this may be grown long before any other cultivated crop can succeed. Perhaps as meadow and pasture land it will be found more profitable than in any other condition, because of the ease with which it may be brought under irrigation, and kept as a water meadow; for there are scarcely any lands better situated for this purpose, or that can be more cheaply and profitably managed in this way, than such lands as are here under consideration.

The cost of such a process of reclamation, as is described in this chapter, will of course depend considerably upon the size of the tract operated upon; the more or less favorable circumstances attending the operation; and the skill with which the works are managed. The most reasonable estimate, when every thing is favorable, is $25 per acre, and from this sum up to $100 per acre may be held to be the probable limits of the cost, unless some very unfavorable circumstances present themselves.

Finally, it may be stated that to insure success in any of the methods of reclamation here considered,

First, the space to be reclaimed must exist within the influence of water which contains much alluvial matter, whether it be situated upon the banks of an inland stream, or of a tidal river or estuary.

Second, that the spaces to be reclaimed shall be allowed to receive the deposit left by the water for as long a period as possible, and the water should not be excluded until, by gradual accretion, the surface of the land has been brought, if possible, to the level of high water of ordinary tides, or above the ordinary level of the stream.

Third, that careful surveys and observations should be made of the amount and quality of the solid matter brought down by the stream, in order to determine the length of time that will be required to complete the reclamation; its cost when complete, and the probable value of the land when it is made and brought under cultivation.

As an instance of the profitable reclamation of marsh lands, bordering upon rivers, and that are periodically overflowed, the "tule" lands of California may be cited. These lands have been formed by gradual accretions, brought down by the rivers, until they have risen above the level of low water. At seasons of flood, these lands are overflowed. When embanked, drained, and reclaimed, these lands bear enormous crops of alfalfa, grass, or wheat. Eight tons of hay, and 40 to 75 bushels of wheat, per acre, have been grown upon these reclaimed lands, and there are none more valuable than these in the whole State. The process of reclamation consists of embanking, draining, and irrigating, although from the moist character of these lands, and the great depth of soil, it is only in the more than usually dry seasons that irrigation is found necessary, and then not by any means to so great an extent as is needed by the valley lands.

INDEX.

* *Illustrated.*

www.ingramcontent.com/pod-product-compliance
Lightning Source LLC
LaVergne TN
LVHW010241110826
845151LV00004B/1361

9781425523428